New Vistas in Transatlantic Science and Technology Cooperation

CHARLES W. WESSNER, *Editor*

Based on a conference
held June 8–9, 1998, in Washington, D.C.

Board on Science, Technology, and Economic Policy

National Research Council

NATIONAL ACADEMY PRESS
Washington, D.C.

NATIONAL ACADEMY PRESS • 2101 Constitution Avenue, N.W. • Washington, D.C. 20418

The National Academy of Sciences is a private, nonprofit society of distinguished scholars engaged in scientific and engineering research, dedicated to the furtherance of science and technology and to their use for the general welfare. Upon the authority of the charter granted to it by Congress in 1863, the Academy has a mandate that requires it to advise the federal government on science and technical matters. Dr. Bruce Alberts is president of the National Academy of Sciences.

The National Academy of Engineering was established in 1964, under the charter of the National Academy of Sciences, as a parallel organization of outstanding engineers. It is autonomous in its administration and in the selection of its members, sharing with the National Academy of Sciences the responsibility for advising the federal government. The National Academy of Engineering also sponsors engineering programs aimed at meeting national needs, encourages education and research, and recognizes the superior achievements of engineers. Dr. William A. Wulf is president of the National Academy of Engineering.

The Institute of Medicine was established in 1970 by the National Academy of Sciences to secure the services of eminent members of appropriate professions in the examination of policy matters pertaining to the health of the public. The Institute acts under the responsibility given to the National Academy of Sciences by its congressional charter to be an adviser to the federal government and, upon its own initiative, to identify issues of medical care, research, and education. Dr. Kenneth I. Shine is president of the Institute of Medicine.

The National Research Council was organized by the National Academy of Sciences in 1916 to associate the broad community of science and technology with the Academy's purposes of furthering knowledge and advising the federal government. Functioning in accordance with general policies determined by the Academy, the Council has become the principal operating agency of both the National Academy of Sciences and the National Academy of Engineering in providing services to the government, the public, and the scientific and engineering communities. The Council is administered jointly by both Academies and the Institute of Medicine. Dr. Bruce Alberts and Dr. William A. Wulf are chairman and vice chairman, respectively, of the National Research Council.

Any opinions, findings, conclusions, or recommendations expressed in this publication are those of the author(s) and do not necessarily reflect the views of the organizations or agencies that provide support for the project.

This work relates to Department of Navy grant N00014-98-1-0762 issued by the Office of Naval Research. The United States Government has a royalty-free license throughout the world in all copyrightable material contained herein.

Limited copies are available from:
Board on Science, Technology,
  and Economic Policy
National Research Council
1055 Thomas Jefferson, N.W.
Washington, D.C. 20007
202-334-2200

Additional copies are available for sale from:
National Academy Press
Box 285
2101 Constitution Ave., N.W.
Washington, D.C. 20055
800-624-6242
202-334-3313 (in the Washington Metropolitan Area)
www.nap.edu

International Standard Book Number 0-309-06197-0

Copyright 1999 by the National Academy of Sciences. All rights reserved.

Printed in the United States of America

For the National Research Council (NRC), this project was overseen by the Board on Science, Technology and Economic Policy (STEP), a standing board of the NRC established by the National Academies of Sciences and Engineering and the Institute of Medicine in 1991. The mandate of the STEP Board is to integrate understanding of scientific, technological, and economic elements in the formulation of national policies to promote the economic well-being of the United States. A distinctive characteristic of STEP's approach is its frequent interactions with public and private sector decisionmakers. STEP bridges the disciplines of business management, engineering, economics, and the social sciences to bring diverse expertise to bear on pressing public policy questions. The members of the STEP Board* and the NRC staff are listed below:

Dale Jorgenson, *Chair*
Frederic Eaton Abbe
Professor of Economics
Harvard University
Cambridge, Massachusetts

James F. Gibbons
Professor of Engineering
Stanford University
Stanford, California

George N. Hatsopoulos
President, Chief Executive Officer
Thermo Electron Corporation
Waltham, Massachusetts

Ralph Landau
Consulting Professor of Economics
Stanford University
Stanford, California

James T. Lynn
Adviser
Lazard Freres
Bethesda, Maryland

Burton John McMurtry
General Partner
Technology Venture Investors
Menlo Park, California

Mark B. Myers
Senior Vice President
Xerox Corporation
Stamford, Connecticut

Ruben Mettler
Chairman and Chief Executive Officer (ret.)
TRW, Inc.
Los Angeles, California

*As of June, 1998.

William J. Spencer, *Vice-Chair*
Chairman, SEMATECH
Austin, Texas

James M. Poterba
Professor of Economics
Massachusetts Institute of Technology
Cambridge, Massachusetts

A. Michael Spence
Dean, Graduate School of Business
Stanford University
Stanford, California

Joseph E. Stiglitz
Senior Vice-President for Development Economics
The World Bank
Washington, D.C.

Alan Wm. Wolff
Managing Partner
Dewey Ballantine
Washington, D.C.

Staff

Stephen A. Merrill
Executive Director

Charles W. Wessner
Program Director

John B. Horrigan
Consultant

Craig M. Schultz
Program Associate

John Oldfield
Program Associate

Laura T. Holliday
Program Associate

STEERING COMMITTEE FOR GOVERNMENT-INDUSTRY PARTNERSHIPS FOR THE DEVELOPMENT OF NEW TECHNOLOGIES

Gordon Moore, *Chair*
Chairman Emeritus
Intel Corporation

M. Kathy Behrens
Robertson Stephen Venture Capital
and STEP Board

Gordon Binder
Chief Executive Officer
Amgen, Inc.

Michael Borrus
Co-Director
Berkeley Roundtable on International
  Economics

Iain Cockburn
Professor of Commerce
  and Business Administration
University of British Columbia

Kenneth Flamm*
Dean Rusk Chair in International Affairs
LBJ School of Public Affairs
University of Texas at Austin

James F. Gibbons
Professor of Engineering
Stanford University
  and STEP Board

W. Clark McFadden
Partner
Dewey Ballantine

Burton John McMurtry
General Partner
Technology Ventures
  and STEP Board

William J. Spencer, V*ice-Chair*
Chairman, SEMATECH
  and STEP Board

Mark B. Myers
Senior Vice President
Xerox Corporation
  and STEP Board

Richard Nelson
George Blumenthal Professor
Columbia University

Charles Trimble
Vice Chairman
Trimble Navigation

John P. Walker
Chairman and Chief Executive Officer
Axys Pharmaceuticals, Inc.

Patrick Windham
Adjunct Professor
Stanford University
Consultant
Science and Technology Policy

Project Staff

Charles W. Wessner
Study Director

John B. Horrigan
Consultant

Ryan L. Catteau
Program Associate

Craig M. Schultz
Program Associate

Laura T. Holliday
Program Associate

---

*At the time of this conference, Dr. Flamm was a Senior Fellow at the Brookings Institution.

NATIONAL RESEARCH COUNCIL
BOARD ON SCIENCE, TECHNOLOGY, AND ECONOMIC POLICY

Conference on New Vistas in Transatlantic
Science and Technology Cooperation

Sponsors

The National Research Council gratefully acknowledges the support of the following sponsors:

The United Kingdom,
(Presidency of the European Union)

The European Commission

Fogarty Center
National Institutes of Health

National Center for Toxicological Research
Food and Drug Administration

Office of Naval Research
United States Navy

U.S. Department of Transportation

National Oceanic and Atmospheric Administration

National Institute of Standards and Technology

Airbus Industrie of North America, Inc.

Chemical Manufacturers Association

Nokia Telecommunications, Inc.

Procter and Gamble

Siemens Corporation

Silicon Valley Group, Inc.

Any opinions expressed in this conference are those of the participants and do not necessarily reflect the views of the project sponsors.

Contents

Preface xiii

I. Introduction 1

II. Proceedings

    Welcome 11
        Kenneth Shine, President, Institute of Medicine

    Opening Remarks 14
        Stuart Eizenstat, Under Secretary of State for Economic,
          Business, and Agricultural Affairs
        Hugo Paemen, Ambassador, European Commission

    Plenary Session I: Trends in Science and Technology Policy

    The U.S. Perspective: The Here and Now Versus the Ideal 20
        Joseph Bordogna, Deputy Director, National Science Foundation

    The EU Perspective on Transatlantic Cooperation 24
        Jorma Routti, Director General DGXII, European Commission

**Plenary Session II: International R&D Cooperation
The EC and U.S. Approach and the Agreement on Science and
Technology Cooperation**

The U.S. Approach to the U.S.-EU S&T Agreement 28
    Melinda Kimble, Acting Assistant Secretary for Oceans and
    International Environmental and Scientific Affairs,
    U.S. Department of State
**The European Union Vista in Transatlantic Science and
Technology Cooperation** 32
    Rainer Gerold, Director, European Commission
**Complementarity of Bilateral and EC Cooperation
with the U.S.** 38
    Paolo Fasella, Director General for Research, Italy

Presentations of Discussions in Breakout Sessions

Group A: Information Technologies 42
    Ray Kammer, National Institute of Standards and
    Technology, Department of Commerce

Group B: Transportation Challenges for the 21st Century 44
    John C. Horsely, Department of Transportation

**Group C: Climate Prediction, Forecasting Applications,
and Impacts** 46
    John Krebs, National Environmental Research Council

**Group D: Human Environmental Health Sciences: Endocrine
Disruptors** 48
    Paul Foster, Chemical Industry Institute of Toxicology

Evening Session

Opening Remarks 51
    John Cadogan, Director General, Research Councils of the
    United Kingdom, for the UK Presidency of the European Union

Keynote Address 54
    Gordon Moore, Chairman Emeritus, Intel Corporation

Second Day's Welcome 60
    William Wulf, President, National Academy of Engineering

Best Practices in Small-Business Technology Development Programs 62
    Moderator: *Helmut List, Chairman, Industrial Research and Development Advisory Council, Austria*

    Industry-Laboratory Cooperation: The Amtex Experiment 63
        Jerry Cogan, Milliken Research

    Laboratory Partnerships with Industry 64
        Dan Hartley, Sandia National Laboratories

    The U.S. Experience with the Small Business Innovation Research Program 67
        Joshua Lerner, Harvard Business School

    The EU Experience with Small- and Medium-Sized Enterprise Development 71
        Patrice Laget, European Commission

    Discussants:
        Jon Baron, U.S. Department of Defense, SBIR Program
        Attilio Stajano, DGIII, European Commission

R&D in the Framework of the New Transatlantic Agenda 76
    Moderator: *Kenneth Flamm, Brookings Institution*

    The 300-mm International Initiative 78
        William Spencer, SEMATECH

    Discussants:
        John Shamaly, Silicon Valley Group, Inc.
        Robert Hance, Motorola
        Michael Borrus, University of California at Berkeley
    Discussion

Internationalization of the Technical Workforce and Transatlantic Cooperation in R&D 88
    Moderator: *William Wulf, President, National Academy of Engineering*

Discussants:
    E. Praestgaard, European Science and Technology Assembly,
      Denmark
    H. Glatz, DaimlerBenz, Germany
    Henri Conze, Ministry for Defense (1993–1996), France
    Gary Poehlein, National Science Foundation
    Dieter Seitzer, Fraunhofer Institute, Erlangen-Nürnberg, Germany
Discussion

Concluding Remarks 94
    Jorma Routti, Director General DGXII, European Commission

Selected Bibliography 98

ANNEX

European Union Research Programs 103
    Professor Jorma Routti and Dr. William Cannell, DGXII,
    European Commission

Multilingual Information Management 117
    Gary Strong, National Science Foundation

Charge for Electronic Commerce Subgroup 122
    Ray Kammer, Director, National Institutes of Standards and Technology

White Papers on Transportation Research 125
    U.S. Department of Transportation

    Opportunities for EU and U.S. Cooperation in Global
    Navigation and Applications

    Intelligent Transportation Systems: Surface Applications

    Intelligent Transportation Systems: Maritime Safety

    Strategic Enabling Research

    Intermodal Transportation: Intermodal Transportation Networks

Agreement for Scientific and Technological Cooperation Between the European Community and the Government of the United States of America 136

Conference Participants 146

Preface

The successful conclusion of the US-EU Agreement on Science and Technology Cooperation offers the prospect of a new chapter in transatlantic cooperation.[1] As with any international agreement in science and technology, the accord's full potential will be realized only if it can encourage mutually beneficial cooperation. With this in mind, responsible officials of the European Union (EU) and the U.S. government contacted the National Research Council's Board on Science, Technology, and Economic Policy (STEP) to discuss how this negotiating success might be publicized and productively exploited. It was agreed that the STEP Board should organize a conference to celebrate the accord, inform the U.S. and European research communities of the agreement, and explore specific opportunities for enhanced cooperation. At the same time, the conference would provide the occasion to review existing and evolving areas of transatlantic cooperation in science and technology from the perception of the United States, the European Commission, and the member states of the European Union.

Given the strong interest and support on both sides of the Atlantic for a major conference, the STEP Board welcomed the opportunity to hold a conference celebrating and advancing transatlantic science and technology cooperation. Encouraging such international cooperation is of great importance to the National Research Council (NRC). Under the leadership of Dr. Bruce Alberts, Dr. William Wulf, and Dr. Kenneth Shine, the NRC has emphasized the role of international

---

[1] For the full text of the agreement, formally known as the Agreement for Scientific and Technological Cooperation Between the European Community and the Government of the United States of America, see the Annex.

cooperation in the advancement of science and human welfare.[2] The Academy therefore was pleased to host an event to encourage cooperation in this domain among nations as we prepare to address the challenges of the 21st Century.

International cooperation is also a central element of a major project now underway under the aegis of the STEP Board. The project focuses on the cooperative activities or partnerships among government, industry, and universities for the development of new technologies. It is being carried out under the direction of a distinguished steering committee, led by Gordon Moore, the Chairman Emeritus of Intel, and is to review the goals and operation of a number of U.S. cooperative programs.[3] These include U.S. programs such as the multi-agency Small Business and Innovative Research program, the Advanced Technology Program of the National Institute of Standards and Technology, and industry partnerships with national laboratories.[4] The project also plans to assess government-industry cooperation in sectors such as biotechnology and computing. The project's ultimate goals are to improve policy makers' understanding of the opportunities and challenges inherent in such partnerships and to make recommendations for best practice, for both international and domestic cooperation.

The recent signing of the agreement and the continued expansion of transatlantic cooperation provided an ideal opportunity for the STEP Board's project entitled *Government-Industry Partnerships for the Development of New Technologies* to explore current international cooperation with Europe, which, along with the United States, is one of the premier centers of global scientific activity. Moreover, U.S.-European cooperation is unparalleled in its scope and depth. It includes expanding opportunities for cooperation at the level of the European Union and vibrant bilateral cooperation among the European member states and the United States. In both regions, public-private collaboration is increasing, raising a rich set of crosscutting policy issues of direct relevance to the STEP Board's work and to the international community as a whole. The high-level policy interest evident on both sides of the Atlantic suggested that the signing of the accord presented a valuable opportunity for the Academy to contribute directly to enhanced transatlantic cooperation.

---

[2]Reflecting the interest in having science more effectively incorporated into U.S. foreign policy, the National Research Council (NRC) is carrying out a study for the Department of State on Science, Technology, Health Issues, and U.S. Foreign Policy.

[3]The STEP Steering Committee members responsible for overseeing the activities associated with this conference are listed in the front matter.

[4]Cooperative Research and Development Agreements, or CRADAs, became a significant element of U.S. technology policy in the 1990s, serving as the principal vehicle for industry-laboratory cooperation.

BACKGROUND

The specific suggestion for the NRC to host a conference to highlight the S&T Agreement, which was then in the final stages of negotiation, emerged through a series of meetings between STEP staff and European Union representatives in Washington and Brussels in 1997. Subsequently, both the Commission and the relevant U.S. government interagency working group endorsed the proposal and requested that the STEP Board hold an event to publicize the agreement and to identify promising areas of potential collaboration. The decision to convene the conference was taken under the leadership of the United Kingdom, which held the EU presidency in the first half of 1998.

THE NEED FOR SUSTAINED EFFORT—
THE CONFERENCE IN EUROPE

From the outset, it was recognized that the range of existing and potential S&T activities is so broad that one conference, no matter how large, would not be sufficient. The partners recognized that the transatlantic S&T relationship is one that could benefit from a sustained effort to share views, review current activities, explore new opportunities, and deepen mutual understanding of the S&T systems in operation on both sides of the Atlantic. Accordingly, it was agreed that a second meeting would be held in June 1999 in Stuttgart, Germany. Major conferences such as these have the advantage of not only assessing current progress, but also of offering a means of focusing the attention of the scientific community on the opportunities presented by expanding transatlantic S&T cooperation.

A SHARED COMMITMENT

Conferences such as the event recorded in this volume do not take place without leadership and commitment. In this regard, the Academy wishes to recognize the leadership and early support of the United Kingdom, in particular, Chris Whaley, the Science Counselor, and Phillipa Rogers, the Attaché for Science and Technology, of the British Embassy in Washington. The leadership and commitment of Dr. Jorma Routti and Dr. Rainer Gerold from the European Commission were essential, as was the encouragement of the Commission's able representative in Washington, Ambassador Hugo Paeman. The STEP Board would like to express a special debt of gratitude to Counselors Patrice Laget and Pablo Amor, of the Delegation of the European Commission. Without their initiative, enthusiasm, good judgment, and support, the Conference could not have taken place, and certainly not within six months of the decision to proceed. On the American side, the STEP Board is grateful for the early encouragement from Dr. Neal Lane, Assistant to the President for Science and Technology and then Director of the National Science Foundation. However, the enthusiasm and financial support of Ray Kammer, Director of the National Institute of Standards and Technology, and Deputy Secretary of Transportation, Mortimer Downey, and his colleague, Fenton Carey, were instrumental in enabling STEP to organize the conference, as was the early interest and support of the National Oceanic and Atmospheric Administration (NOAA). The STEP Board wishes to acknowledge the Chief of Naval Research, Admiral Paul G. Gaffney, for his contribution at a crucial juncture in the preparation of the conference.

A number of distinguished individuals deserve recognition for their willingness to review this report. These individuals were chosen for their diverse perspectives and technical expertise, in accordance with procedures approved by the National Research Council's Report Review Committee. The purpose of this independent review is to provide candid and critical comments that will assist the institution in making the published report as sound as possible and to ensure that the report meets institutional standards for objectivity, evidence, and responsiveness to the study charge. The review comments and draft manuscript remain confidential to protect the integrity of the deliberative process. We wish to thank the following individuals for their participation in the review process: Dr. Gerald Dinneen, the review coordinator, Dr. Albert N. Link, Professor of Economics, University of North Carolina, Greensboro, Dr. David Bruce Audretsch, Director and Ameritech Professor of Economic Development, Indiana University, Dr. Richard Thayer, President, Telecommunications and Technologies International, and Dr. John Boright, Executive Director, Office of International Affairs at the National Research Council. Although these individuals have provided constructive comments and suggestions, it must be emphasized that responsibility for the final content of this report rests entirely with the STEP Board and the NRC. It should also be emphasized that, although the conference participants identified numerous areas of potential collaboration, no formal recommendations are made by the National Research Council and the National Academies of Science and Engineering.

Many of the topics covered in the conference are of interest to industry as well as to the research community. As a result, the conference benefited from the endorsement of Siemens Corporation, Procter and Gamble, Airbus Industries, Nokia Telecommunications, Silicon Valley Group, and the Chemical Manufacturers Association. Without their interest, confidence, and support, the conference could not have been organized in the time frame and on the scale required. Last, the STEP Board would like to thank Thomas Kalil, of the White House National Economic Council, for his leadership in identifying topics of common interest and outstanding participants, as well as for his participation. Among the STEP staff, Dr. John Horrigan and John Oldfield deserve recognition for their commitment, skill, and energy in organizing STEP's largest conference to date. Dr. Horrigan also played an instrumental role in producing the conference report. As the acknowledgments above suggest, the organization of this conference was a cooperative effort benefiting from the genuine interest, on both sides of the Atlantic, in developing transatlantic S&T cooperation.

The evident mutual interest, indeed enthusiasm, for transatlantic cooperation in science and technology does not, however, mean that there are no challenges to overcome. Effective cooperation requires that we recognize, but are not deterred by differences in perspectives and practices. Addressing these issues and identifying common ground, while sharing the burdens and the fruits of research, are the essence of sustainable international cooperation. Our goal in hosting this conference was to contribute, with many others, to a new and productive chapter in transatlantic cooperation.

Charles W. Wessner

Introduction

Collaboration in science and technology is a hallmark of the late twentieth century. Cooperation across national frontiers is expanding, aided by new communications technologies and motivated by the global nature of many scientific challenges. Within nations, cooperation between government and industry is expanding to meet national goals and common technological challenges. Strategic alliances among businesses also are expanding dramatically, enabling firms to meet a variety of goals, from sharing expertise and costs to establishing global standards and ensuring market access for final products. The global nature of many technological challenges and the enormous expense associated with developing new technologies have made international cooperation an essential element of national science and technology policy.

Successful collaboration, among both nations and firms, requires care and commitment. Care is necessary because much depends on the choice of partner and the clear articulation of goals and responsibilities. The sustained commitment of individuals and institutions and, not least, sustained funding, are essential for cooperative activities to bear fruit. To fully realize the benefits of international collaboration, substantial vision and commitment are required of researchers and policy makers alike.

The December 1997 signature of the Agreement for Scientific and Technological Cooperation Between the European Community and the Government of the United States of America (US-EU S&T Agreement)[1] represented a significant achievement, creating a bridge between the research and development (R&D)

---

[1] See the Annex for the text of the Agreement.

systems on both sides of the Atlantic. To ensure that the Agreement had an immediate and positive effect on transatlantic cooperation, both parties agreed to hold a major conference during the British presidency of the European Union on June 8–9, 1998. Organized under the auspices of the National Research Council's Board on Science, Technology, and Economic Policy, the conference had three objectives: First, it served as an important opportunity to publicize the Agreement within the research community and among policy makers on both sides of the Atlantic. Second, the conference brought together experts in substantive areas where opportunities for fruitful collaboration were believed to exist. Third, the conference sought to examine crosscutting issues of common interest in areas such as the framework for R&D collaboration, small business development, and the internationalization of the technical workforce.

To achieve these objectives, the conference deliberations were organized around three broad categories:

- **the policy context,** articulated by senior officials;
- **specific research topics** discussed in small meetings of researchers and policymakers interested in collaboration and
- **crosscutting issues** of broad interest to the two communities.

THE POLICY CONTEXT FOR THE U.S.-EU S&T AGREEMENT

As Under Secretary of State Stuart Eizenstat noted in his opening remarks, the US-EU S&T Agreement is part of the New Transatlantic Agenda of 1995, which seeks to strengthen ties between the European Union and the United States. The European Commission's Ambassador to the United States, Hugo Paemen, noted that the new S&T agreement complements both the New Transatlantic Agenda and the EU's Fifth Framework Programme. The Ambassador also observed that the Agreement has taken effect at a time when the United States has launched its 21st Century Research Fund. This confluence of events makes the Agreement especially timely.

The Deputy Director of the National Science Foundation, Joseph Bordogna, reminded the audience that scientific and technological advance must take place with human consequences in mind. Dr. Bordogna urged the gathering of scientists and policy makers to use the new U.S.-EU S&T Agreement to address human needs within the context of a global imperative to reduce inequality and protect the environment. Professor Jorma Routti, Director General of DG XII, highlighted the socioeconomic dimension of the Fifth Framework Programme by laying out some of its specifics. This Programme will focus on the life sciences and biotechnology, information technologies, competitive and sustainable growth of industries, as well as energy and the environment. International cooperation is a key element in the Programme, as are promoting innovation in small businesses and improving worker training.

Acting Assistant Secretary of State, Melinda Kimble, focused her remarks on implementation of the Agreement. Article 6 of the Agreement calls for convening a joint consultative group (JCG) to discuss implementation. The informal JCG meeting, scheduled immediately following the conference, addressed many of the topics taken up by the conference. These included endocrine disrupters, information science and technology, materials research, intermodal transportation and intelligent transportation systems, measurement equivalents, health and environmental effects of radiation, and climate-change prediction. Ms. Kimble emphasized that a cornerstone of the strategy to realize the U.S.-EU S&T Agreement's potential will be to enlist the public and private sectors in joint efforts to implement and publicize the Agreement.

In offering his perspective on the Agreement, Rainier Gerold, Director General of the European Commission, stressed that the U.S.-EU S&T Agreement is particularly important to Europe in light of the growing importance of the Fifth Framework Programme to the overall European research and development (R&D) enterprise. R&D to be funded by the Fifth Framework Programme totals 3.5 billion ecu. Although this is three times larger than the First Framework Programme in real terms, it amounts to only 5 percent of all S&T funding among EU members. However, the Framework Programme is increasingly important because it addresses strategic issues, such as health and the environment. Because the policy challenges are global in nature, the payoffs from cooperation can be great. The New Transatlantic Agenda's section on R&D recognizes this reality and explicitly calls for wider cooperation between the United States and European Union in science and technology.

Notwithstanding the widespread recognition of the benefits of greater transatlantic cooperation, Dr. Gerold recalled the challenges the negotiators faced in reaching the Agreement. For example, he noted the difficulties encountered in engaging separate R&D systems and the specific differences that emerged on intellectual property and rules for foreign participation. Dr. Gerold concluded that the new S&T Agreement will work best if it mobilizes researchers and stimulates bottom-up cooperation on both sides of the Atlantic.

Italy's Director General for Research, Dr. Paolo Fasella, provided the perspective of a leading European member state on the accord. Dr. Fasella reminded the audience of the current importance and likely continued prominence of bilateral cooperation between the United States and individual European countries. For example, the large European intergovernmental research organization for particle physics, CERN, operates outside the framework of the S&T Agreement and will undoubtedly continue to be the focal point for cooperation on this type of research. Citing an example from his own country, Dr. Fasella pointed to the Agreement between the U.S. National Aeronautics and Space Administration and the Italian Space Agency to work together to develop technology for the International Space Station. As valuable as the new S&T Aagreement is, Dr. Fasella emphasized that bilateral cooperation will continue to play an important and complementary role in the future.

POTENTIAL AREAS FOR COLLABORATION

The afternoon session of the first day consisted of breakout sessions on each of the four topical areas: information technologies, transportation, climate prediction (forecasting applications and impacts), and human environmental health sciences (endocrine disrupters). More than one topic was covered in these areas, and the assembled scientific and policy specialists spent two hours exploring potential areas of transatlantic cooperation. After the breakout sessions, session chairs reported their recommendations to all conference participants.

For information technology, Ray Kammer, Director of the National Institutes of Standards and Technology, reported on electronic commerce, cross-lingual information management, and the Next Generation Internet. In general, each group identified specific issues in each field in which the United States and the European Union might collaborate. An important issue, particularly for electronic commerce and cross-lingual information management, will be standards setting. Global standards are inherently transnational. Agreement on standards is essential if we are to realize the full potential of many new information technologies. The groups also urged that a process be put in place to facilitate collaboration between the United States and the European Union in information technologies.

U.S. Associate Deputy Secretary of Transportation John Horsely reported on the three topics covered by the breakout sessions on transportation: intermodal transportation, intelligent transportation systems, and strategic enabling research. Mr. Horsely noted that an important goal of transportation research is to encourage sustainable and competitive growth, while working to decouple traffic growth from economic growth. The transportation conferees identified specific common research interests and suggested convening a workshop on the institutional impediments to transatlantic collaboration on transportation research.

John Krebs, a member of the United Kingdom's National Environmental Research Council, summarized the breakout session on climate prediction, forecasting applications and impacts. Dr. Krebs noted that the climate research community is already very well connected internationally beause of the global nature of climate research. Any mechanism established under the U.S.-EU S&T Agreement must therefore be transparent and more convenient than existing mechanisms. The climate group cautioned strongly against "double jeopardy" in an application process that requires researchers to clear bureaucratic hurdles in both Brussels and Washington. The climate breakout group identified a number of areas for potential collaboration, such as the terrestrial environment and the precision of climate forecasting, encouraged further dialogue, and suggested that representatives of the business and polar science communities be included in future meetings.

Paul Foster of the Chemical Industry Institute of Toxicology summarized the breakout session on human environmental health science, which focused on endocrine disrupters. These are chemicals released into the environment that can function as hormones. Some believe that they may have a serious impact on humans, wildlife, and vegetation. Breakout participants identified risk assessment and further examination of the fundamental biology of endocrine disrupters as possible areas for collaborative research. The breakout group suggested that the procedure for obtaining funds under the Agreement be made clear to the research community and that a panel of experts be gathered to flesh out additional collaborative opportunities. The group also raised the prospect of trade issues arising from endocrine disrupters, because government intervention into the management of chemicals easily could have trade consequences.

KEYNOTE SPEAKERS

At the end of the first day, the conference heard from John Cadogan, Director General, Research Councils of the United Kingdom, and, for the evening's keynote address, from Gordon Moore, Chairman Emeritus of Intel. Interestingly, both speakers placed innovation in the broader context of the advancement of knowledge, and both emphasized the need to ensure adequate support for curiosity-driven research.

Dr. Cadogan pointed out that the creativity of scientists in the laboratory must be harnessed in order to make collaboration between the United States and the European Union effective. A conference is not enough to realize the potential of the Agreement; it is necessary to cultivate the innovators in the laboratory. Dr. Cadogan observed that in Europe, scientists have become skilled in collaboration, but must do better at making new discoveries. In the same way that the latter half of this century was transformed by the electronics revolution, he expects that biotechnology will prove the most fruitful ground for discovery in the next century.

Dr. Moore presented an overview of the semiconductor industry's history of innovation, based on his personal experience with Intel, the leading U.S. semiconductor producer. He highlighted the role that international cooperation could play in tomorrow's innovation. Dr. Moore explained that, from its earliest days, when the transistor was invented at Bell Laboratories, the semiconductor industry has been a beneficiary of the industrial research system. From their modest beginnings, semiconductors have become a pervasive part of every industry. The semiconductor industry now has over $150 billion in annual revenues and supports a larger electronics industry of $1 trillion worldwide. The ability to pack more and more processing capability on the same piece of silicon has led to the spectacular cost declines and functionality improvements that have been the hallmark of the industry. A few years after the invention of the semiconductor in 1948, it was estimated that the cost of making a single transistor soon would be 60 cents. Today, a single transistor costs less than one-eighth of a microdollar, or 120 nanobucks.[2]

The system that brought about such innovations is, however, under stress. Large industrial labs have been downsized in recent years and the R&D that many companies conduct is increasingly short term in nature. Long-term research is less frequently undertaken, not only because it is costly and uncertain, but also because companies find it difficult to fully capture its benefits.

Yet the need for long-term research in the semiconductor industry remains as urgent as ever. Optical lithography is approaching the physical limits of its ability to etch circuit lines onto silicon.[3] Alternative technologies are under development, but the research is expensive and it will take several years for promising alternatives to come into production. For example, x- ray lithography shows promise, but there are technical and financial challenges to overcome. To address these common challenges, the semiconductor industry established the National Technology Roadmap for Semiconductors as a mechanism to identify technical challenges and coordinate industry, government, and university research.

Partly as a result of the road-map approach, funding shortfalls for some basic research increasingly are recognized. Dr. Moore explained that it generally is accepted that it is most appropriate for government to support widely applicable university research, noting that "even if the research fails, you still get trained students." For this reason, the semiconductor industry has created a number of programs designed to support university research, but additional government support for university research would be a worthwhile investment. In conclusion, Dr. Moore noted that international collaboration is bound to be valuable in meeting future research challenges, especially among firms and universities.

CROSSCUTTING ISSUES

Developing Small Business

The second day of the conference focused on additional topics of interest to the technological and economic development of Europe and the United States. Recognizing the importance of entrepreneurship to ensuring a technologically dynamic economy, conferees discussed their efforts to promote small business development. One case drawn from the U.S. experience focused on the Ameri-

---

[2]A microdollar is one millionth (10^{-6}) of a dollar and a nanobuck is one billionth (10^{-9}) of a dollar.

[3]At one time, it was thought that optical lithography would be able to manufacture chips whose line-widths were no smaller than 250 nanometers (nm). However, existing optical techniques have proven capable of reaching the 100nm line-width level.

can Textile Partnership. This is an effort by the entire U.S. textile industry in cooperation with the national laboratories to promote the use of advanced technology in textile manufacturing. The U.S. national laboratories develop a wide range of technologies to meet national defense missions, and many of these technologies can have applications in industry. Having defense laboratories and industry work together on common technological challenges is often a way for both parties to gain expertise while increasing taxpayers' return on their investment in the national laboratories.

Another example of U.S. efforts to promote the small business sector is the Small Business Innovation Research (SBIR) program. This program, established by Congress in 1982, sets aside a fixed percentage of selected U.S. agencies' R&D budgets for grants to small businesses. The program's purpose, as laid out in its authorizing legislation, is to augment private-sector commercialization of publicly funded R&D, increase small business participation in federal R&D programs, and improve the dissemination of federal R&D opportunities, particularly to small businesses owned by women or socially and economically disadvantaged individuals.[4] In working toward those goals, the government hopes to take greater advantage of the innovative capacities of small businesses. Recent research suggests that this program has registered some success for national research programs. An empirical analysis presented by Professor Joshua Lerner of the Harvard Business School showed that recipients of SBIR awards perform better than similar small businesses that have not received SBIR awards.[5]

The European Commission has programs with similar goals. Dr. Patrice Laget of the European Commission's delegation to Washington described the EU's efforts to increase collaboration among small- and medium-sized enterprises (SMEs) throughout Europe. Some success has been achieved: From the Third Framework Programme to the Fourth, the number of SMEs participating in EU research programs doubled. It is anticipated that SME participation will continue to grow in the Fifth Framework Programme.

International Cooperation on Semiconductors

Turning to broader issues of transatlantic R&D cooperation, the conference heard presentations on several existing international cooperative R&D enterprises. One major new international initiative, called the I300I, is developing next-generation tools for the semiconductor industry to manufacture chips on 300-mm

---

[4]For legislative background on SBIR, see Robert B. Archibald and David H. Finifter, "Perspectives on the evaluation of the SBIR program with an application to the NASA-Langley research center." Paper presented at the American Economic Association meetings, December, 1998.

[5]Joshua Lerner. 1999. "Public venture capital: Rationales and evaluation." In National Research Council, *The SBIR Program: Challenges and Opportunities.* Washington, D.C.: National Academy Press.

silicon wafers, as opposed to the current 200-mm standard. Organized as a subsidiary of the U.S. semiconductor consortium SEMATECH, the I300I initiative is open to international membership. It now includes firms from Europe, Korea, and Taiwan. Japan has also launched a 300-mm conversion project, called Silicon Leading Edge Technology (SELETE), all of whose shareholders are Japanese firms.[6] Whether national or international in orientation, consortia are necessary to meet the staggering cost of industrywide conversion to 300 mm, which may reach $20 billion. Unlike some previous conversions, this cost is well beyond the means of any single firm.

Similar cooperative efforts are under way in Europe. For example, the High Epsilon Materials Cluster Optimized Rapid Deposition project complements the 300-mm conversion project, albeit on a smaller scale. This project ensures that appropriate materials for 300-mm manufacturing will be available when the entire 300-mm tool set is ready. Funded at $11 million annually, the project has European and American membership.

Internationalization of the Technical Workforce

The third panel discussed the internationalization of the technical workforce, a topic of growing importance on both sides of the Atlantic. This panel emphasized improving exchanges of students between the United States and Europe. As one European observed, many European scientists receive their training in the United States, but there has been a drop-off of U.S. students seeking training or postdoctorate positions in Europe. Encouraging further ties, whether through professor-to-professor contacts or more formal programs, will require the interest of individuals and the availability of funding. U.S. scientists-in-training must be encouraged to view international experience as beneficial to their career, and it was suggested that business in the United States promote such internationalization. For transatlantic collaboration to work, individuals must be engaged; for it to endure, engagement of individuals early in their career is paramount.

Effective collaboration is rarely an easy task; individuals, institutions, and countries naturally have different perspectives and objectives in carrying out research. Nevertheless, the global character of society's most pressing problems and the rising cost of addressing them, provide great incentives for international collaboration. Fortunately, the explosive growth in communications made possible by information technology will facilitate transatlantic collaboration. Although face-to-face meetings are often essential in the early stages of business

---

[6]For an excellent analysis of I300I and SELETE, see Rose Marie Ham, Greg Linden, and Melissa Appleyard, "The evolving role of semiconductor consortia in the U.S. and Japan," *California Management Review*, Vol. 41, No. 1, Fall 1998. pp. 137–163. Korea's Samsung Electronics is listed as a participant in SELETE, but it is not a shareholder.

and academic relationships, advanced information networks are making long-distance professional collaboration a daily occurrence. The combination of expanded personal contacts and today's information technologies can move transatlantic cooperation to a new level.

For individuals and their institutions to work together effectively, leadership at the highest levels is required. To develop and maintain such leadership, and thereby realize the enormous benefits possible from European and American collaboration, a policy framework that accepts institutional and philosophical differences is needed. The new S&T Agreement holds the promise of such a framework, offering the United States and Europe the opportunity to jointly address the challenges of the twenty-first century.

Charles W. Wessner

Welcome

Kenneth Shine
President, Institute of Medicine

On behalf of the National Academy of Sciences it is my pleasure to welcome you to our two-day conference on transatlantic science and technology (S&T) cooperation. As you know, the United States and European Union (EU) signed the U.S.-EU Science and Technology Agreement on December 5, 1997. We gather here to celebrate the agreement and to explore ways to build cooperation between the United States and the European Community and its member nations.

We are very pleased that so many of you are able to join us here this morning. As we get under way I want to extend a special welcome to our European friends, who traveled so far to be with us today. Let me recognize in particular:

- John Cadogan, Director General of the Research Councils of the United Kingdom, who is here as representative of the U.K. Presidency of the European Union;
- Jorma Routti, Director General for Science, Research, and Development in the European Commission;
- Paolo Fasella, Italy's Director General for Research; and
- Hugo Paemen, Ambassador of the European Commission here in Washington, who is well known to all as the able representative of the European Community.

We also have a distinguished group of participants from the U.S. government whom I am pleased to recognize and thank for their participation today:

- Mortimer Downey, Deputy Secretary of Transportation, whose leadership and energy are well known to the Academy;

- Stuart Eizenstat, Under Secretary of State, who in his previous capacity as ambassador to the European Commission was instrumental in initiating this agreement;
- Melinda Kimble, Assistant Secretary of State for Oceans and International Environmental and Scientific Affairs;
- Ray Kammer, Director of the National Institute of Standards and Technology. We owe a special debt to Ray for his early support of the conference and for his interest in transatlantic cooperation in science and technology.
- And last but by no means least, Joseph Bordogna, Deputy Director of the National Science Foundation and a leading figure in U.S. science policy.

The S&T agreement that we celebrate today is a major accomplishment and having a conference of this scale is an extremely promising beginning. It reflects, of course, like any good conference, the combination of interesting topics and people with the energy and commitment to make it work. Conferences do not just happen; they are put together with the cooperation of many people who are actively engaged in other professional activities.

It is for this reason that, on behalf of the National Research Council, I want to extend our sincerest thanks to those of you who have contributed intellectually and financially to this conference. I especially want to thank the European representatives here in Washington: Phillipa Rogers of the British Embassy and the remarkably well-informed Patrice Laget and Pablo Amor of the European Commission delegation. While there are too many U.S. government representatives to thank personally, I would be remiss not to mention Admiral Gaffney of the U.S. Navy, whose interest in and support of the work of the Academy are deeply appreciated.

We are here today to bring the agreement to life—to infuse it with meaning so that the real work of S&T cooperation can get under way. First, we must recognize that there is ample scope for cooperation between the United States and the European Union. Indeed, given the transatlantic nature of this conference, it is perhaps worth underscoring just how much the United States and European Union spend on research and development (R&D). We all have our problems, but our economies are the drivers of science and innovation in today's world. For example:

- In 1995 U.S. firms spent $132 billion on industrial R&D, with $24 billion of that funded by the government. This was a 10 percent increase from 1994.
- While the United States leads the world in spending on industrial R&D, the gap between the United States and the European Union is declining.

- In 1973 about 52 percent of the world's industrial R&D was performed by the United States, with 28 percent performed by the European Community.
- By 1995 the U.S. share had fallen to 46 percent, while the European Community's had risen to 30 percent.

Together the United States and the European Union account for around three-quarters of the world's industrial R&D. This gives us a tremendous opportunity for mutually beneficial cooperation in science and technology. But we must choose carefully the areas in which we engage in cooperative S&T development. It is important to find the areas that are most likely to yield productive collaboration and concentrate our energies there. Then we must carefully manage the cooperative process, settling on well-defined goals, specifying timetables, and committing resources to create some of the world's best science and technology.

This, by the way, is one of the best features of our conference. We are here to discuss and exchange our views regarding broader S&T policy objectives, but, importantly, we are not limited to that dimension. The conference is also designed to examine cutting-edge science and technology across a remarkably broad range of topics. These issues are of direct concern not only to our economies but also to the health and welfare of our citizens. Indeed, personally, I am pleased that the conference recognizes that we have more at stake than mutual gain for our economies. We have a *responsibility* to use our talents, our education, and our public and private resources to leave the world better than we found it.

Communications technologies make the world a seemingly smaller place all the time. We can harness communications and computing technologies to create safe, efficient, and clean means of transportation. The continuing revolution in health sciences and biotechnology opens up new possibilities for protecting and maintaining human health. And as the fruits of science and technology create greater abundance for more people on our planet, we must also use science and technology to protect our planet from environmental degradation.

As you can see from the agenda, we have two full and fascinating days ahead of us. We have the challenge in the near term to create a climate where transatlantic S&T cooperation can flourish. With hard work and dedication, I am confident we can meet the challenge.

In the long term we have the equally formidable challenges of, first, sustaining cooperation and, second, harnessing science and technology for the greater good of the economies and societies of the United States and Europe. We must turn to each of these challenges with dedication and the optimism that our talents are equal to the tasks ahead. I wish you the very best, not only today and tomorrow but also for a new vista in transatlantic cooperation.

Opening Remarks

Stuart Eizenstat
Under Secretary of State for Economic,
Business, and Agricultural Affairs

INTRODUCTION

Thank you for inviting me here today. I want to congratulate the National Research Council for organizing this conference and bringing together such an impressive assemblage of American and European scientists and researchers, technical experts, and policymakers. We are counting on all of you to help us turn the opportunities created by the new U.S.-EU Science and Technology (S&T) Agreement into realities. You represent the "cutting edge" in your respective fields, and through this conference we hope that ideas are shared and cooperation enhanced to the mutual benefit of both the United States and the European Union.

The launching of our new S&T agreement is an occasion of great expectation as well as tremendous personal and professional satisfaction. I have watched and encouraged the emergence of this agreement from three different vantage points, beginning with my service as Ambassador to the EU, then as Under Secretary of Commerce, and now as Under Secretary of State for Economic, Business, and Agricultural Affairs. In fact, the signing of this agreement is the realization of the commitment I personally made as Ambassador to the EU in 1995. A mandate between myself and Mrs. Edith Cresson, commissioner responsible for research, innovation, education, training, and youth, was obtained to conclude an agreement by 1997. We fulfilled that mandate and promise at the U.S.-EU summit in December 1997.

S&T AGREEMENT

The agreement will serve as a broad framework for cooperation, enabling some of our most distinguished scientists and best research institutions to collaborate on a wider range of scientific endeavors and to initiate new joint programs. In addition, the agreement establishes a common ground for handling the allocation and protection of intellectual property rights resulting from joint research. Throughout this process I have been convinced that the United States and the EU had an important stake in expanding our scientific and technological collaboration.

First, it is imperative for our scientific and technical communities to work together in this era of globalization. Second, it is critical for our economic and trade interests to develop agreed intellectual property rules, common databases, and mutually acceptable standards. Third, it sends an important political signal that building stronger transatlantic bridges is in both our interests.

The agreement, which is based on the principles of mutual benefit, reciprocal opportunities for cooperation, and equitable and fair treatment, should help researchers and institutions on both sides of the Atlantic, including subsidiaries of both European and American companies, work more closely in a wide variety of research areas. While we have enjoyed extensive S&T cooperation with Europe and many EU member states for a long time, this agreement opens new areas previously closed to mutually beneficial cooperative activities and provides protection for intellectual property rights. The agreement encourages cooperation in areas where the United States and the EU are doing some of the most advanced research in the world: environment, agriculture, information and communications technologies, biomedicine, health, and manufacturing processes.

NEW TRANSATLANTIC AGENDA

This agreement is a solid example of the enhanced cooperation between the United States and the EU. President Clinton has consistently made clear his personal commitment to stronger transatlantic ties, and this commitment was manifested in the New Transatlantic Agenda (NTA). Agreed upon with the EU at the December 1995 summit, the NTA has strengthened and enhanced our partnership. It broadens our cooperation and has the most complete set of cooperative mechanisms we have ever had: semiannual summit meetings, regular subcabinet-level meetings, and a broad range of working-level contacts.

Most importantly, the NTA provides a blueprint for strengthening cooperation between the United States and Europe into the twenty-first century. The NTA recognizes that ours is a constantly developing relationship that must adapt to internal and external changes, and changes brought about by science and technology are a fundamental part of our relationship.

The S&T agreement is a key instrument for advancing the NTA goal of expanding U.S.-European scientific cooperation across the Atlantic. The agreement is further concrete evidence of the U.S. State Department's "firm commitment to international S&T" highlighted by my colleague, Under Secretary of

State for Political Affairs Tom Pickering, in a recent speech to the American Association for the Advancement of Science. As Ambassador Pickering said, we are all working hard "to make science relevant to our foreign policy and to bring the department in close touch with the opportunities presented to us in foreign affairs of a closer relationship to the underworld of science and technology."

Acting Assistant Secretary Melinda Kimble and the State Department's Bureau of Oceans and International Environmental and Scientific Affairs are devoting additional resources to ensure the successful implementation of this agreement and are reaching out to technical agencies and the scientific community for new ideas and help. They are also taking full advantage of the Internet and other available vehicles to ensure widespread dissemination of information about the agreement.

There is much still to be done. We are eager to hear fresh and imaginative ideas on how we can seize the new opportunities before us. Good luck in your discussions over the next two days. You all have an opportunity to help direct what we fully expect to be an extremely productive relationship. And I urge you to take full advantage of it! At the State Department, we await your suggestions and input. We will work closely with our European Commission partners to ensure that we live up to the promise of our new agreement. Thank you for inviting me to join you this morning and I look forward to receiving and reviewing your report.

Opening Remarks

Hugo Paemen
Ambassador, European Commission

It is a pleasure to be here today for what I hope will be a stimulating conference on the theme of transatlantic science and technology (S&T) cooperation. Fifty-one years ago, U.S. Secretary of State George Marshall set forth a vision of a democratic Europe that would be whole, free, and at peace. His hope was that such a Europe could act as a full and equal partner of the United States. Much of Europe responded, laying the groundwork for the European Union. Since then, our transatlantic partnership has been the leading force for peace, democracy, prosperity, and development for ourselves and the world.

The New Transatlantic Agenda signed in Madrid in December 1995 confirms the commitment of the European Union and the United States to further develop, in the new global economic and geopolitical environment, our common goal of fostering an active and vibrant transatlantic community. This should be done by deepening and broadening the political and economic ties that bind us, as well as the social, cultural, educational, and, last but not least, scientific ties. The conclusion last December of the agreement between the European Community and the United States for a program of cooperation in science and technology clearly fits the overall goals that I mentioned.

This conference could not come at a better time. At the same moment that the European Commission is finalizing the Fifth Framework Programme for European Research and Training activities covering the period 1998 to 2002, the United States has launched the 21st Century Research Fund, also a five-year initiative, to enable the various S&T agencies to focus more intensely on the president's goals for science and technology. Let us use this conference to inject into both of these major efforts the transatlantic dimension that our S&T agreement has called for.

Good collaborative links between public agencies and academic and industrial researchers are clearly essential if the high quality of our academic research in Europe and the United States is to be matched by our ability to produce high-tech goods, processes, and services that can be winners in the global marketplace. This is why we thought there would be considerable benefit, particularly for those of you who are involved in putting together and implementing research and development (R&D) and innovation programs, in sharing experiences by bringing together administrators and industrial and academic researchers to exchange ideas on good practice. I do not say "best practice" in this context because what works well in one country may work less well elsewhere. So I hope that you will leave Washington, D.C., in two days with plenty of good practical ideas to consider and a range of new contacts—people you may wish to stay in touch with in the future—and which will be sorely needed if the U.S.-EU S&T agreement is to become the success we all expect.

I would like to share with you a couple of concrete messages that seem to me important not to lose sight of in the course of these two days. The first one is the understanding on the European side of the complementary role of the European Community research programs that provide the necessary European dimension to the member states' national R&D programs. I am sure that Professors Routti and Fasella will develop this key point in much more detail during their presentations. Please keep in mind then that this is a meeting about finding new ways to make the European R&D program, built on the success of the respective national programs, identify joint research opportunities with our American counterparts.

This European dimension has many expressions, but perhaps a concrete and easily understood one is the requirement to assemble research consortia in which at least two European organizations are involved while recognizing at the same time that many U.S. federally funded programs involve only one organization.

The second thought I would like to share with you concerns the increasing importance of collaborative partnerships between industry, academia, and national labs as a way to speed up the ability of our economies to shorten the product-to-market time frame and increase competitiveness in global markets. The very first European Framework Programme stressed these partnerships. This was done not only from the conviction that this was the right way to increase the competitiveness of the European industry but also and especially out of necessity. The European Union's R&D funds represent a very small percentage of the funds that member states allocate to R&D. Forging partnerships has been a way to leverage these funds and provide the "biggest bang for the buck." Additionally, the growing complexity of research necessitates a multidisciplinary approach in which all interested actors, industry, academia, and research labs provide input and participate actively.

On the U.S. side, we also observe a growing recognition of the importance of partnerships to tackle difficult research areas. Examples include the Partnership for the Next-Generation Vehicle and high-performance computing networks. In my opinion, the time is ripe to extend this thinking beyond European Union and U.S. borders and provide a transatlantic dimension to these partnerships.

In order for these partnerships to work, they need to be built using a "bottom-up" approach. By this I mean that agreements such as our recently signed S&T one only constitute an institutional framework. Without the active support of the scientific community, they become empty documents. An open and fluid dialogue among the practitioners will jump-start the process and provide a momentum that can never be there by simply putting together administrators in charge of research on both sides. This conference endeavors to provide such a climate for at least four research areas by bringing together investigators and managers active on both sides.

I believe the task ahead of you during these two days is to work within these boundary constraints, as mathematicians like to call them, and to try and identify proposals for future collaborative work. Such work should not only make sense from a scientific and engineering perspective but should also fit the requirements and missions of the various U.S. federal research agencies as well as the European Commission. Of course, the task will not be completed in two days, and a similar conference is planned for early 1999 to address additional domains of possible collaboration.

This conference is predicated on the premise that no one has a monopoly on wisdom. I hope that by the end of tomorrow all of us will have learned something new of value that we can apply in our own context. If we achieve that, we will have made a genuine contribution to improving the transatlantic scientific dialogue. I wish you well in your deliberations.

Trends in Science and Technology Policy

The U.S. Perspective: The Here and Now Versus the Ideal

Joseph Bordogna
Deputy Director, National Science Foundation

I am very pleased to participate in today's discussion of transatlantic science and technology (S&T) cooperation. The last century is replete with transatlantic collaborations of every nature from national security alliances to matters as diverse as public policy and political elections, scientific exchanges, and arts and cultural events.

This plenary session on trends in S&T policy is part of all of our nations' long tradition of bridging the broad Atlantic Ocean with good communication and good ideas. This conference to inaugurate cooperation under the new U.S.-European Union (EU) science and technology agreement extends that tradition into the twenty-first century.

My assigned topic for this talk is "Trends in Science and Technology Policy—The U.S. Perspective." While I speak to this, my comments will be focused primarily on what those trends should ideally be rather than what they necessarily are now. The title of my remarks is "Trends in Science and Technology Policy: The Here and Now Versus the Ideal."

Many of us here today are civil servants in the broadest and most generous sense of that term. We serve as officials of public institutions that often tend toward bureaucratic-sclerosis over time. Our first task, it seems to me, should be to commit ourselves to proving *incorrect* the blessedly unknown scholar who said, "Bureaucracy defends the status quo long past the time when the quo has lost its status." It is a humorous but not untrue commentary on the danger of institutions holding onto the past instead of lifting their sights to the future. Our task is to recognize and retain what is valuable from the past while envisioning a future based on inevitable change.

I should add that, although the definition of bureaucracy refers primarily to the public sector, bureaucrats have been known to exist in the private sector, too. The difference is often their briefer period of survival. We are fortunate indeed to have the able participation and advice of some of industry's best talent with us for this conference.

Let me begin with a comment by paleontologist and evolutionary biologist Stephen Jay Gould from his recent essay in *Science* magazine (Feb. 6, 1998). He said, "... science cannot be separated from political change, if only because the primary motor of social reorganization throughout human history, from the advent of agriculture to the acme of modern industry, has been fueled by ... scientific knowledge." Although scientist and nonscientist alike can marvel at the power of our knowledge in science and technology, it is the intersection of this knowledge with the goals and needs of society that is our larger responsibility.

From the first time humans left the confines of this planet to venture into space close to four decades ago, the limited circle of our globe and the even tighter circle of our dependency on each other have become increasingly apparent. Those first photographs of earth taken from space spoke not only of our shape and size in the universe but of our unity. We are all citizens of the small blue planet. And on this planet the advancement of civilization has, in many respects, been driven by the scientific and engineering research of each succeeding generation.

We can all agree that science is a force absolutely fundamental to our well-being and, in fact, survival. Indeed, science and society are interdependent. There is an inextricable relationship between the diverse science, engineering, and technology activities in all our nations and the public policy efforts that enable populations, economies, and nations to reap maximum benefit from advances in knowledge and understanding.

Although we know this connection by both instinct and example, we are only slowly coming to the recognition that science and technology, and its concomitant policy, must be seriously concerned with the many and great unsolved problems of humankind. This latter premise moves our planning and projections to another, quite different, level.

I do not in any way lightly dismiss the consistent increase in science, engineering, and technological knowledge that moves across national borders. Neither do I discount the widening net of international collaborations, not only among our nations but with all nations. These are positive and contributory trends. But none of us can escape the contradiction in contemporary society that we are able to do increasingly outstanding science at the same time that many societal disparities and problems are increasing. Those of us in the S&T policy community are in a unique position to address these issues. The deliberations of this very meeting can establish, for the record, a distinction between the current trends in S&T policy and the ideal trends for the very same.

Although many in the science and engineering community may not think of these matters as their individual responsibility, one of history's most eminent scientists spoke of this very issue decades ago. In 1931, before World War II and in the deepest days of economic depression, Einstein admonished the science community in an address at the California Institute of Technology. He said, "Concern for man himself and his fate must always form the chief interest of all technical endeavors, concern for the great unsolved problems of the organization of labor and the distribution of goods—in order that the creations of our mind shall be a blessing and not a curse to mankind. Never forget this in the midst of your diagrams and equations."

Here we are 67 years later finally building consensus for his wisdom. Einstein takes us back to our fundamental values as guidance—our concern for humanity and its fate. I believe that it is in those terms that we must work toward the more ideal trends in S&T policy in the twenty-first century.

Since the end of the Cold War in 1989, the era of East-West rivalry has been eclipsed by an emerging era of North-South realities and relationships. President Clinton's trip to Africa exemplifies this recognition. Nuclear testing by India and Pakistan also is part of that new reality. This emerging era comes with new challenges, interdependent consequences, shared international responsibilities, and mutual opportunities. Much of the opportunity will be powered by the world science and engineering community.

There is a global imperative to close the widening gap between the haves and have-nots—not through handouts or handdowns but through building knowledge and capacity in poorer nations to enable them to create their own wealth. Although America is thought of as a rich industrial nation we are facing a similarly widening division in our own borders. Many of your nations are experiencing similar phenomena. The gap between rich and poor and skilled and unskilled in our nations or elsewhere in the world cannot bode well for our collective future.

In 1960 the world population was 3 billion. We all know that by the turn of the century that number will double to 6 billion. This will have occurred in less than four decades. Most of the world's population growth and much of its economic expansion will occur in the Southern Hemisphere. Here too will exist the potential for the deepest problems of hunger, poverty, and disease, as well as for energy supply, vast environmental devastation and their incumbent emergencies.

Although the 130 plus developing countries already account for four-fifths of the world's people, they only account for one-sixth of its economic output. This pervasive condition of poverty devastates individuals as well as nations and has far-reaching implications for all of the world's citizens and nations. Poverty degrades the dignity of us all as human beings no matter where it occurs, North, South, East, or West.

It is clear that Einstein would have us be mindful to think not only of saving our planet for future generations but of saving the planet's current generation. Our reverence for humanity's habitat must include a reverence and compassion for humanity itself. Our only hope of saving either rests in a commitment to save both. Sustainable development cannot mean sustaining poverty in those places where it exists.

The major problems facing the whole global society are human problems. And they will require more than technical solutions. These problems emerge out of complex patterns of overlapping consequences. For example, over the past several decades, the investment that industrial nations have made in improved nutrition, medical technologies, and public health have all coalesced to boost life expectancy in Europe and the United States from less than 47 years in 1890 to 75.5 years in 1993. Japan has done even better. More recently, this trend is also emerging in developing countries.[1] This is surely an advance to celebrate for all humanity.

However, as this life expectancy trend increases, nations will struggle to support their elderly populations with a decreasing proportion of their populations of wage-earning ages. Thus, our triumph of better health and longer life will also pose an economic dilemma. Our job will be to create opportunity from this and other impending dilemmas.

We cannot deny that there are overlapping consequences of poverty, planetary devastation, illiteracy, aging populations, communicable diseases, mass migrations of immigrants, agricultural output, energy supply, and others. Grappling with these issues collectively might seem like a completely unmanageable task, at best. But we do not have the luxury of making choices. We do have new technological tools for innovative approaches. We can, indeed, make the same leaps of majestic proportion that created every other milestone of human progress.

We know that energy, environment, and economics form the triple challenge of the coming century; they are inextricably wedded. We know that despite national and cultural differences, every nation—big or small, rich or impoverished, agricultural or industrial or postindustrial (as some speculate), democratic or dictatorial—each is woven into the interlaced fabric, some would say a postindustrial digital fabric, of the world's economy and ecology.

We may be gathered today to contemplate future collaborations among our several nations and through the European Union, but our vision must necessarily encompass a far broader concern. These discussions are transatlantic by association, but our genuine universe of thought must be transglobal if we are to move from the "here and now in science and technology trends toward the ideal."

I wish you every success in defining areas not only for transatlantic cooperation but for global vision as well. Thank you.

---

[1] *Science* Vol. 273, pp. 46–48. July 5, 1996.

Trends in Science and Technology Policy

The EU Perspective on Transatlantic Cooperation

Jorma Routti
Director General DGXII
European Commission

I welcome the opportunity to be here today to explore the opportunities associated with the new U.S.-European Union (EU) science and technology (S&T) agreement. We believe it has rich potential, and we are grateful to the National Academy of Sciences for hosting this event; after all, a key ingredient in cooperation is knowing your partner. Accordingly, my charge today is to say something about trends in S&T policy in the European Union, and to do this I will focus on new themes in the EU's Fifth Framework Programme for research and development. Perhaps the most prominent theme in the Fifth Framework Programme is the socioeconomic dimension of research and development (R&D). Moving into the next century, the Fifth Framework Programme hopes to promote R&D that improves the quality of life of all of our citizens while doing so in an environmentally sustainable way. This involves a interdisciplinary approach to R&D, as we work as a community to push the frontiers of science and technology and, where appropriate, with partners across the Atlantic and around the globe.

But before getting into the details of how we hope to accomplish these goals, let me tell you why we fund collaborative R&D at the level of the European Community and provide some background on past Framework Programmes.

BENEFITS OF EUROPEAN COLLABORATIVE RESEARCH

In Europe we recognize that we must invest more in research and technology. European Union countries spend 1.8 percent of GDP (gross domestic product) on civil R&D, as opposed to 2.5 percent in the United States and 2.8 percent in Japan. There are several reasons why community-wide collaboration in R&D makes sense:

- Bringing together researchers from several EU countries deepens the pool of research talent in any one project and broadens linkages across borders, which in turn contributes to R&D dynamism.
- An increasing number of research issues, such as climate change and marine and terrestrial ecosystems, can be carried out effectively only with trans-national coordination.
- Large-scale research infrastructure is increasing costly, even when distributed across many EU members.

THE NATURE AND EVOLUTION OF FRAMEWORK PROGRAMMES

The EU's Framework Programme comprises four activities:

- research, technology development, and demonstration, which make up 87 percent of all expenditures in the Fourth Framework Programme.
- international cooperation in research, which involves partnerships with non-EU countries and international organizations, builds R&D links to less developed countries and fosters access of EU countries to cutting-edge research elsewhere;
- dissemination and exploitation of research, through technology transfer and monitoring of best R&D practices elsewhere; and
- stimulation of training of researchers through international fellowship programs.

The First Framework Programme was established in 1984 and initially was modest in scale; today, the Fourth Framework Programme is three times the size in real terms of the first program and amounts to 3.5 billion Ecu annually. The Framework Programme accounts for 4 percent of all civil R&D in the EU and 4 percent of the total European Community budget. In general, the EU's four Framework Programmes have funded R&D in five areas: energy, life sciences, environment, industrial materials and technologies, and information and communications technologies. Research priorities have shifted over the years. Energy has diminished in relative importance over time, and information and communications technologies, having peaked in funding in the 1980s, have declined somewhat. Areas such as transportation and socioeconomic research have, in contrast, experienced funding increases.

THE FIFTH FRAMEWORK PROGRAMME: A NEW STRATEGIC APPROACH

The Fifth Framework Programme, which runs from 1998 to 2002, recognizes that, with the EU rapidly integrating, a broader strategy based on knowledge, innovation, and education and training is necessary. Society faces major issues such as environmental protection and well-being of citizens, in addition to economic competitiveness. To meet these challenges, *concentration* and *flexibility* have become the key concepts in the Fifth Framework Programme. This means that the EU plans to become more flexible in the allocation of resources while focusing on socioeconomic aspects of research and its application, not just technology.

To adhere to our principles of concentration and flexibility, the Fifth Framework Programme is organized into four thematic programs and three horizontal programmes, with a budget of 16.3 Ecu over four years. This contrasts with the Fourth Framework Programme, which involved some 20 separate and specific research programs. The four thematic areas are life sciences and biotechnology, user-friendly information technologies, competitive and sustainable growth of industries, and energy and environment.

To maximize returns, the objective is to concentrate on a limited number of objectives in order to strengthen the EU's S&T base. This involves a focus on generic R&D and support of the research infrastructure throughout the European Community. In our planned research on information technologies, we seek to develop next-generation digital services that improve citizens' access to government services and deliberation. It also means developing multimedia content and tools that enable cultural and linguistic diversity while encouraging electronic publishing.

For sustainable growth our objective is to improve land transport and marine technologies in order to move people and goods more efficiently. Our plan also calls for improvement in air traffic control technologies. (For a detailed list of thematic areas, I recommend the paper I have with me today on the Fifth Framework Program, coauthored with William Cannell; see Appendix).

The three horizontal programs, which are designed to complement the thematic ones, are international cooperation, promotion of innovation and participation of small and medium enterprises (SMEs), and improving training and mobility of researchers.

If we refer to the agenda for this conference, we see how nicely the thematic areas in the Fifth Framework Programme fit. Our general topic here is international cooperation, and we have separate sessions tomorrow on SMEs and the internationalization of the technical work force.

THE SOCIAL DIMENSION

Let me conclude by underscoring an important theme from the Fifth Framework Programme—socio-economic research. We do maintain our focus on the natural sciences and technology, but we have a new emphasis on the socioeconomic dimensions of science and technology. As technology becomes a more pervasive part of all our lives, it is important to acknowledge how social, behavioral, and economic factors work together to shape the development and applications of technology. By recognizing these interrelated factors the European Union wants to make science and technology do more for the quality of life of its citizens, while leaving a healthy planet for our children and grandchildren. We have a broad range of topics for potential cooperation that we will explore today and tomorrow. I look forward both to the discussions on substantive topics and the expanded cooperation that will result.

International R&D Cooperation

The U.S. Approach to the U.S.-EU S&T Agreement

Melinda Kimble
Acting Assistant Secretary for Oceans and
International Environmental and Scientific Affairs
U.S. Department of State

Thank you for the opportunity to participate in this special event—an announcement and celebration of the signing of the U.S.-EU S&T agreement and the kickoff of the implementation phase. As Under Secretary Eizenstat has mentioned, the signing of the agreement was a fulfillment of a commitment made by Presidents Santer and Clinton in 1995 when they signed the New Transatlantic Agenda. After two and a half years of negotiations and six months of preparation for the joint consultative group meeting, the United States and the European Union have a right to celebrate. We also have a responsibility to pursue implementation expeditiously in order for the United States to realize the agreement's full potential for this transatlantic partnership. This gathering should add impulse and insight to the task. I commend our hard-working delegates on both sides, the National Academy of Sciences, and all participants to engage actively as we embark on this mission of cooperation.

I will speak briefly about our strategy for implementing the agreement and touch on the State Department's approach to science and technology, in general, a topic my bureau, the Bureau of Oceans and International Environmental and Scientific Affairs (OES), has been studying seriously for some time.

The State Department, and specifically OES, is the custodian for some three dozen framework S&T agreements and hundreds of MOUs (memoranda of understanding), with more arriving each day, that form the basis of bilateral S&T cooperation worldwide. We take this role very seriously. For example, in the Western European area this past year alone, we have had important and successful bilateral review meetings or consultations with Finland, Italy, Spain, and Portugal. These activities will continue since there are activities in these arrangements that are more appropriately performed at the member-state level. Sir Leon

Brittan recognized the compatibility of the U.S.-EU S&T agreement with other bilateral agreements at the December 5, 1997, signing ceremony, underscoring that the S&T agreement shall not impinge on or prejudice existing bilateral agreements with the United States but rather would complement them.

That said, the U.S.-EU S&T agreement could well become the largest of all S&T agreements owing to its potential and scope, as well as the billions of dollars in R&D involved. Moreover, cooperation between the world's best scientists on cutting-edge research will accrue enormous savings by avoiding duplicative efforts and will yield significant beneficial breakthroughs for the entire world. The agreement also provides for the protection of intellectual property rights—an essential means of encouraging research and technological innovation.

The U.S. government's approach to the S&T agreement can be summed up in one word: proactive. As the "executive agent" responsible for liaisoning with the European Commission and the catalyst for energizing over 15 U.S. government agencies to support the negotiation and implementation of the agreement, the State Department, specifically OES, is working hard to make the agreement operational. To exploit the momentum of the December 1997 signing, our strategy calls for:

- the early convening of the joint consultative group (JCG) called for under Article 6 to jointly chart next steps in the implementation process;
- the designation of priority areas for cooperation, which include four or five items under the "sectors for cooperative activities" found under Article 4(a) of the agreement; and
- the publication and promotion of the agreement through our public affairs apparatus, including the posting of the agreement, joint statement, points of contact, and other information on the OES website: www.state.gov/www/global/oes.

The first meeting of the JCG, to be cochaired by Professor Routti and myself will be held June 10 at the State Department. It will be an "informal" meeting of the JCG since the EC must still ratify the agreement. Barring any unforeseen circumstances, we plan to hold the first official JCG in Brussels on October 21.

The topics chosen for the informal JCG may sound familiar to those familiar with the agenda for this conference: endocrine disruptors, information science and technology, materials research, intermodal transportation and intelligent transportation systems, measurement equivalents, health and environmental effects of radiation, and climate change prediction. These priority areas reflect our agencies' interest in cooperative projects (some actually have draft MOUs or other implementing arrangement documents ready to go); the importance and timeliness of these topics; and, after close consultation with our EU colleagues, a mutual acceptance of the appropriateness of joint projects on a priority basis in these areas. I can also report that we are working to identify several other priority areas for joint cooperation for the October JCG, which will possibly include renewable energy resources, biomedicine and health, telematics, and agriculture.

In terms of publicizing and promoting the agreement—the third element of our strategy—we and our European partners have tried to make this agreement and its implementation a public/private enterprise to the extent possible. We realize much of the joint research will be done by government agencies in cooperation with private scientific laboratories, academic institutions, consortia, and small and medium enterprises. We also understand the interest of the business community in the agreement and what it portends for future R&D trends. There exists an interesting dynamic here, where we look to private industry as leaders in technological advancement and engines for progress. I would imagine that they, in turn, are interested in our policy directions—and to ensure we complement and reinforce their plans. This conference, part celebration and part mutual edification, will help all of us clarify our priorities and learn from our shared experiences regarding the potentials and pitfalls of cooperation between two very different, some might say incompatible, systems.

Turning now to a more macro view, I would like to say a few words on the topic of science at the State Department. I am acutely aware of the criticism leveled at us over the past year from the perception that the department is deemphasizing the S&T function both here and abroad. Quite honestly, there were legitimate grounds for these concerns, as in some cases mandatory downsizing claimed its share of EST positions. The EST cone in the foreign service has become subsumed once again into the economic cone. This all occurred against a dramatic 84 percent increase in multilateral environmental negotiations. At a time when the EUR bureau was adding observers to Bosnia, OES was forced to sacrifice the routine for the urgent.

It is axiomatic to say that science undergirds all we are trying to do on the environmental side. Our climate change, toxic waste, and biosafety talks may get the attention and headlines, but without the science and technology that come from it there can be no appreciation for the magnitude of the problems or a plan for confronting them. With the climate change issue, for example, it was the consensus of 2,000 scientists from around the world that anthropogenic factors affected the world's climate; scientific and economic models which will allow us to map out a cost-effective strategy to meet the challenge; and scientific methods and technological innovations to monitor and contribute to a global reduction of greenhouse gas emissions.

Besides our efforts in launching this impressive agreement with the EU, which we are commemorating this week, and maintaining the other bilateral S&T relationships around the world, allow me to mention our recent efforts and plans to bolster the science function at the State Department. As part of the department's environmental diplomacy initiative (inaugurated by Secretary Christopher and endorsed by Secretary Albright) to mainstream EST issues into U.S. foreign policy, we are establishing a global network of regional EST hubs that will facilitate interaction between the U.S. government and other governments, nongovernmental organizations, and multilateral organizations in a particular region. Our regional hub in Copenhagen, for example, will serve to galvanize and coordinate aspects of our Baltic EST policies with host governments, our embassies, and others in the region.

OES will soon hire a science adviser who will report directly to me. It is envisioned that with this science adviser, the various bureaus in the department that deal with scientific matters, including offices in OES, Political-Military Affairs, Economic and Business Affairs, and others, will come together in an S&T working group or "science team" to coordinate our international programs internally and then on an interagency basis.

The OES 2000 plan is a blueprint for getting the necessary resources to better enact the concepts put forward in the Environmental Diplomacy Initiative. Funding for the regional hubs to hold seminars on emissions trading and joint implementation, to set up education and training centers, and to assist exchanges of information and scientific personnel are all part of the OES 2000 plan. The President's speech on information technology at the Massachusetts Institute of Technology highlighted the importance of computer literacy and glimpsed a vision of a future where access to information and ideas will transform societies and enhance our quality of life. We are positioning ourselves to engage other governments and their scientific communities to initiate or enhance mutually beneficial exchanges.

Under Secretary Thomas Pickering, in his April 30 speech to the American Association for the Advancement of Science (alluded to earlier by Under Secretary Eizenstat), acknowledged the hard choices ahead regarding the allocation of resources for science, technology, and environment at the State Department. He mentioned the review currently being conducted by John Boright of the National Academy of Sciences (and formerly of OES) concerning the role of science at the State Department. We look forward to Dr. Boright's report so that we can better meet the S&T policy challenges of the twenty-first century in order to, as Mr. Pickering put it, "better advance global economic and humanitarian interests . . . [resulting] in more science-based cooperation, a cleaner planet, a healthier world population, regional stability, and global economic growth."

In closing, I would like to thank the Academy for its efforts in putting together this impressive and important event and congratulate Professor Jorma Routti and all the European and U.S. participants for a good start on realizing the potential of productive cooperation under the U.S.-EU S&T agreement.

International R&D Cooperation

The European Union Vista in Transatlantic Science and Technology Cooperation

Rainer Gerold
Director
European Commission

Melinda Kimble has presented the U.S. approach to the U.S.-EU Science and Technology (S&T) Agreement. My task is to present the European Union vista. To help the transition to a discussion afterwards, I shall answer four basic questions. And in order to make sure that there is a discussion, I shall openly address some questions that caused difficulties in the negotiations.

One may ask why an S&T cooperation agreement between the European Union and the United States is necessary. There are after all long-standing and extensive cooperative links between individual researchers, research institutions, and industrial laboratories in Europe and the United States. Indeed a considerable number of European researchers and Research, Technology, and Development (RTD) managers—for example, the previous speaker, Professor Routti, and the following one, Professor Fasella—have spent several years of their careers in U.S. laboratories. In addition, there is a multitude of agreements, both nongovernmental and governmental, between individual EU member states and the United States. Furthermore, centrally managed European Union research accounts for only about 10 percent of the total research effort and 5 percent of the S&T funding in the European Union, the vast majority of it being nationally funded (funding by the EU is only up to 50 percent of total costs).

The answer lies in the fact that the European Union Framework Programme is playing an increasingly important role in addressing the strategic questions our society faces. Professor Routti has just described the activities foreseen for the Framework Programme over the next four years. It is designed to complement the national research activities of the EU member states and to provide a mechanism and funding to allow governmental, academic, and industrial researchers in European member states to work together on questions of general European relevance.

Professor Fasella will elaborate on the complementarity between national and EU S&T efforts in the next presentation. Most of the problems addressed by the Framework Programme—for example, control of infectious diseases, the aging of the population, and global climate change, as well as the challenges of the information society or of mobility—are similar on this side of the Atlantic and must also be faced by American society. There is, therefore, a genuine mutual interest in cooperating to solve them through common effort. Furthermore, the EU and United States both face the challenges of strengthening the partnership between government, universities, and industry and of ensuring that research is interconnected to other policy areas.

Several research agreements involving the EU and the United States already exist—for example, in the fields of nuclear fission, thermonuclear fusion, and biotechnology. We are both partners in a series of multilateral agreements, such as the International Science and Technology Centre in Moscow, dedicated to the conversion of military research capacity in Russia to civilian research demands, or the Agreement on Intelligent Manufacturing Systems, which may receive further incentives from the U.S.-EU S&T agreement. However, these agreements concern only focused topics. The mutual interest of a much wider cooperation was recognized at the highest political level by including a chapter on RTD cooperation in the New Transatlantic Agenda and by specifically requesting a comprehensive S&T cooperation agreement.

The S&T agreement is therefore part and parcel of a much wider political initiative aimed at strengthening the transatlantic partnership as stressed earlier by Under Secretary Eizenstat. In this overall context both sides committed themselves to "foster to the fullest extent practicable the involvement of participants in cooperative activities under this agreement with a view of providing comparable opportunities for participation in their scientific, technological and development activities" subject to applicable laws, regulations, and policies (Article 5a of the S&T agreement). They also agreed on the following basic principles: mutual benefit based on overall balance of advantages, reciprocal opportunities to engage in cooperative activities, and equitable and fair treatment of partners.

So, if the advantage of such a cooperation framework is so obvious, why has it taken two years to conclude the agreement? There are several elements to the answer. First, the RTD management, funding, and policy development systems differ between the two partners. Whereas the U.S. RTD system is highly decentralized, with many different authorities responsible for different scientific areas and each with a particular set of rules, the European RTD Framework Programme, which has developed over 15 years, has a single authority working on the basis of a harmonized set of rules. In the United States, federal RTD support is given in a variety of forms—for example, grants to individual institutes or to multipartner consortia. In the European Union, funding is provided exclusively through open competitive calls for proposals and only to consortia of several (on average five or six) partners from different countries, with preferably a mix from academia, industry, and other users.

Second, there were different views on the arrangements for intellectual property rights and exploitation of results achieved through joint research projects. We all know that there are several thorny intellectual property rights issues pending between the EU and the United States. Some differences also exist between EU countries, which complicates matters and might sometimes give the United States the impression of having to dance with an octopus. The Trans Atlantic Business Dialogue is actively looking into these questions, and we hope that satisfactory solutions will be found in order to overcome obstacles to cooperation.

Third, there has been some debate on both sides of the Atlantic as to whether foreign participation in our respective RTD activities is an asset or a liability. I refer, for example, to a report of the U.S. National Academy of Engineering (1996).

In Europe we have developed the view that international cooperation is very much an asset. The positive experience with the Framework Programme has contributed a great deal to this view. While the first priority of the Framework Programme has always been to stimulate RTD cooperation between EU member countries, there has also been an openness to partners from outside the EU, naturally to other European countries but also to non-European countries. Indeed, promotion of RTD cooperation with third countries is specifically highlighted in Article 130 g (b) of the treaty establishing the European Union Community.

In the participation rules of the upcoming Fifth Framework Programme, different categories of non-EU cooperation partners are envisaged. First are countries that are formally associated with the Framework Programme and that contribute financially to the community research budget. RTD institutes from these countries can participate on a basis similar to the EU member states, including financial support. These countries will include Norway, Iceland, and Liechtenstein. Negotiations for association have been concluded with Israel; they are ongoing with Switzerland and are about to start with the 10 Central and Eastern European countries and Cyprus, which are all candidates for membership in the European Union. This may bring the pool of fully eligible cooperating countries from 15 to about 30 as well as increase the total sum of funding available for cooperative research.

S&T entities from all remaining European countries, including Russia and the Ukraine, are fully eligible as partners in consortia under the Framework Programme. The same is true for S&T entities from the 12 Mediterranean Partner countries of the EU.

Similar conditions apply to all countries with whom the EU has concluded S&T agreements such as Canada, Australia, South Africa, and now the United States. Agreements with Russia, China, and some other countries are envisaged. Scientists from these countries may participate in the calls for proposals under the Framework Programme together with at least two partners from the EU.

Partners from countries not contributing to the budget of the Framework Programme will have to bear their own expenses, but they fully share the results of the research achieved in common. Participation by scientists in all remaining countries of the world is possible but more limited. We are convinced that such openness to cooperation, which may have been a choice in the past, has become a must in today's global society. The European Union is not a fortress seeking to remain self-contained; on the contrary, it seeks cooperation with partners in other parts of the world, including with its major competitors, because in the long run such cooperation is in everyone's interest. Some tasks are just too great or too expensive to be tackled alone.

During the negotiation of the S&T agreement, we have retained the impression that U.S. legislation and practice are more restrictive as regards access to federal S&T programs than the rules of the EU Framework Programme. We are, however, confident that in the implementation of this agreement and in order to make a success of it, the U.S. agencies will arrive at the same conclusion as we have—namely, that such an open attitude toward cooperation is in their own interest. Concretely, this also means that cooperation should not be limited to fields where both sides have previously agreed, in a more or less formal way, to cooperate—that is, the top-down approach. We would expect that the bottom-up approach—that is, cooperation activities proposed spontaneously by scientists, without previous encouragement by the administration—should not only be possible but also welcomed by both sides.

A fourth point that blocked the negotiations for several months concerned the definition of "foreign" participants. This was finally overcome by way of a side letter to the agreement. In the EU all research entities established in the European Union are eligible to participate in the Framework Programme. This means that any subsidiary of an American company based in the European Union (e.g., IBM Europe) is fully eligible to participate in research consortia and to be funded by the Framework Programme. This is not so obvious for subsidiaries of EU companies established in the United States. We see here a certain contradiction with the efforts to promote transatlantic trade and investment. Therefore, we welcome the understanding that both parties to the agreement expect an equitable treatment of their subsidiaries as regards opportunities for participation in S&T activities. We trust that public financing in the United States will not exclude American subsidiaries of European firms.

What is the scope of the agreement and what are the forms of cooperation foreseen? In negotiating the agreement both parties were conscious that nowadays more and more research is tackled using an interdisciplinary approach. Article 4 of the agreement enumerates practically all fields of research. There are only three exceptions: military research, nuclear research (which is governed by a separate agreement), and research related to plant and animal varieties. The latter was excluded in a side letter at the request of the United States. As new harmonized legislation is to be approved by the European Parliament, the partners may want to look again at this exclusion.

Regarding the forms of cooperation, the agreement covers virtually everything from exchange of information and joint seminars to integrated research projects. During the running-in period, the lighter forms of cooperation will prevail. I am confident, however, that with time more and more joint projects will emerge. Hopefully, these will include projects that are conceived together from the beginning and not only projects that are put together after each side has already taken its conceptual decisions. The former are still rare between the United States and Europe.

What remains to be done? Now that the U.S. and EU administrations have put the skeleton in place, it remains for the scientific community to put flesh on the bones by developing joint research activities. It remains for the administration, however, to ensure that the S&T communities on each side of the Atlantic are fully informed of the new possibilities offered by the agreement and that the commitments undertaken on either side are explained. We must bear in mind that information is not the same as mobilization and that the research community must be convinced of the usefulness of cooperation and of the novelty of the possibilities offered by the S&T agreement. We must also take care to avoid placing too many administrative hurdles in the way of the cooperation and encourage scientists to initiate spontaneous proposals in addition to the activities envisaged by the administration.

Let me say here that we do no expect detailed knowledge from any U.S. partner wanting to join a European consortium about the procedures ruling the European S&T programs, they should trust their European partners to solve the administrative questions. A series of "awareness actions" will be undertaken. Indeed, the first of these major awareness events is this conference here today. It is also an important first step in the identification of specific research topics where cooperation could be particularly fruitful.

Having been involved in the negotiations of the agreement from the beginning and having lived through a number of difficulties both next door at the State Department and in Brussels, I am particularly satisfied that this conference has triggered such interest from government, academia, and industry. I am sure that our U.S. counterparts share this feeling, particularly Ron Lorton, who chaired the U.S. team and suffered with us during what seemed to be an endless story. This conference is a promising start, and I would like to thank everyone who contributed to its success.

A corresponding conference is envisaged in Europe next spring, possibly in Germany, to cover additional topics of mutual interest. I certainly hope that U.S. participation in that conference will be as great as the European participation in Washington today.

The second activity aimed at improving awareness of the possibilities offered by the S&T agreement is the establishment of a clear, user-friendly home page on the World Wide Web, available to all interested persons. The EU and, as I understand, the United States have already established such home pages, which must be updated continuously.

A third and very important instrument will be the designation of contact persons for each scientific area, at both the U.S. agencies and the European Commission. These contact persons will be fully briefed on the possibilities and undertakings of each party to the agreement. A list of the EU contacts is available at the conference. The Science Counsellors of the European Commission and of the individual EU member states in Washington and those of the United States in Brussels and in EU countries will also have to contribute substantially to the information campaign.

The joint EU-U.S. Consultative Committee established by the agreement will play a crucial role in providing the necessary drive for the success of the agreement and in solving any problems that may arise. The first informal meeting will be in two days.

Intensification of cooperation will take time; mutual confidence will not be developed overnight. The closer S&T activity is to application and the market, the more reluctance there will be to associate outside partners. There must be an overall balance and both partners must be convinced that cooperation is in their interest. These are general rules, not specific to the EU-U.S. relationship, and they clearly must be respected. Nevertheless, I trust that when we come together in two or three years, there will be a measurable increase in EU-U.S. S&T cooperation.

In Europe we have a long experience in international RTD cooperation between the EU member states. The evolution has been tremendous and has led to a true RTD cooperation culture despite language and cultural differences and the growth of the European Union from 6 to 15 member states. It has strengthened the fabric of European science and innovation and has been so successful that intra-EU RTD cooperation is now no longer even considered "international cooperation." We will be happy to share this experience with the United States. There is a lot to be done so that this joint venture can contribute to the adventure that is science. Let's attack it together!

International R&D Cooperation

Complementarity of Bilateral and EC Cooperation with the U.S.

Paolo Fasella
Director General for Reseach, Italy

My comments are a short sequel to the presentations by Dr. Bordogna and Dr. Routti. I will focus on the bilateral science and technology (S&T) collaboration between the United States and a single European Union member state—Italy—in view of the new situation created by the U.S.-EU agreement. S&T international collaboration is of growing importance for all countries and poses new and diverse problems. The diversity of problems requires diverse solutions. We therefore welcome the new U.S.-EU agreement, and I am personally grateful to the organizers of this conference for having invited me to participate.

THE PRINCIPLE OF SUBSIDIARITY

In choosing which activities to propose for pursuit in the framework of the U.S.-EU agreement and which ones at the bilateral level, the principle of "subsidiarity," widely used in the European Union, is a valid reference. According to this principle, a sort of European research version of the 10th amendment of the U.S. Constitution, the EU framework should be used for those actions that are required by commission policies or which can be carried out more efficiently at the community rather than the national level. I shall give some examples as to how this principle could be applied in the context of bilateral S&T collaboration between the United States and a single member state of the European Union. These collaborations are, in fact, quite important. Italy is a founding member of the European Union and for more than 10 years has benefited from a bilateral agreement with the United States.

The EU-U.S. agreement foresees reciprocity and a balance of benefits for the two partners. These must be pursued in a situation that is de facto nonsymmetric, given the resources available on the EU side. The EU Framework Programme covers only a small percentage of government-supported research in Europe—4 to 5 percent in terms of direct contributions by the European Union. This actually corresponds to a chiffre d'affair (or turnover) that is twice as high, since most EU interventions cover only 50 percent of the costs, with the remaining 50 percent contributed by other sources. The European Union cannot guarantee the access of U.S. partners to national or regional research activities (more than 90 percent of publicly supported research in Europe) over which it has no authority or competence. On the other hand, most of the multibillion dollars of the U.S. federal research budget is managed by powerful and authoritative institutions, such as the National Institutes of Health and the National Science Foundation, which enjoy a large degree of autonomy, including most decisions about international collaboration.

TYPES OF U.S.-EU COLLABORATION

Even with these constraints there are many research actions that can be conceived and implemented jointly by the United States and the European Union on the basis of the new agreement. In large-scale research and development (R&D) ventures, the European Union can be a partner of a size comparable to the United States; this is to the advantage of both partners. This whole meeting is dedicated to the EU-U.S. collaboration, and I shall only confirm here the will of the Italian government that Italy, as a member of the European Union, will participate vigorously in these activities. However, I also want to talk about other forms of S&T collaboration involving the governments of the United States and Italy. They are worth considering because they are complementary to those foreseen by the U.S.-EU agreement.

One form of collaboration concerns large European intergovernmental research institutions. They are generally dedicated to specific branches of science and technology, such as CERN for particle physics, the European Molecular Biology Laboratory, the European Space Agency (ESA), and the European Southern Observatory for astronomy. They were created after World War II as the means to achieve, through collaboration, the level of human and economic resources required by world competition in areas too expensive for single European countries. After some trials and errors, they have been quite successful. CERN has become the world leader in some branches of physics and has attracted the participation in financial as well as scientific terms of the United States and Japan. ESA collaborates and also competes with NASA (National Aeronautics and Space Administration). As we shall see later, this does not exclude bilateral collaboration between the United States and Italy in some space projects. Cooperation by the United States with these and other non-EU European research organizations provides an additional channel for U.S. and Italian researchers to work together and must be kept in mind when planning bilateral U.S.-Italian collaboration.

Another set of opportunities is provided by "wider than Europe" and "worldwide" organizations to which both the United States and Italy belong. The European Union has representatives in many of these organizations. In this case, EU member states, including Italy, can participate in joint activities individually and/or as members of the European Union. Some of these organizations, such as the United Nations and the Organization for Economic Cooperation and Development (OECD) cover a wide range of interests. Others, such as the World Health Organization and the Food and Agriculture Organization, are sectoral. In the OECD it is customary for EU member states to consult generally before meetings, under the chairmanship of the country charged with the pro tempore presidency of the EU Council and with the participation of the European Commission. Other examples of "wider than Europe" activities are the Human Frontiers Science program for molecular biology and neurobiology, ITER (for controlled nuclear fusion), and the International Center for Science and Technology, for the conversion to peaceful purposes of military S&T research in the former Soviet Union. The experience that EU member states have gained in intergovernmental S&T collaboration involving many countries has been valuable, as has been the case with the OECD Megascience Forum.

Membership in the European Union still allows individual member states latitude to undertake internal initiatives. Italy, for instance, launched the International Center for Genetic Engineering and Biotechnology and the Abdus Salam International Center for Theoretical Physics, both in Trieste; U.S. scientists participate in both. Collaboration between the United States and individual EU member states, like Italy, could be further developed within these organizations in research areas such as biosafety, bioethics, biocomputing, global change, and oceanography.

BILATERAL COOPERATION

Besides the above-mentioned multinational framework, the United States and Italy have created a bilateral system. Here I shall mention some activities implemented, or under consideration, at the bilateral level, taking into account the principles of complementarity and subsidiarity.

One example concerns space. An agreement between NASA and ASI (the Italian Space Agency) has recently been signed; the collaboration includes the development of mini pressurized logistical modules for the International Space Station and the successful launch of the Cassini mission to the planet Saturn. The other example concerns biomedical research. As Professor Routti said, collaboration between the United States and the European Union is envisaged in the areas of biomedicine, even though the EU's Fifth Framework Programme does not assign a high priority to research on cardiovascular diseases. This is not because this sector is not considered important. Rather in the spirit of subsidiarity (relevance and adequacy of national programs), it is thought that research on biomedicine should take place on a national basis or, where appropriate, using bilateral mechanisms.

The United States and Italy think that collaboration in this field, and especially in what has been called a "rational approach" to epidemiology, is a very worthwhile development. Under these circumstances, collaboration in this area could be suitably carried out in the framework of the agreement between the United States and Italy. Other research areas, which are widely covered in the European Union's Fifth Framework Programme, such as prevention and control of infectious diseases and biomedical problems related to aging, could be the object of collaborations within the new EU-U.S. agreement.

In conclusion, the very rapid and diversified worldwide development of S&T requires new forms of collaboration. The position of the United States and the European Union in world science and technology is such that the interactions among them are particularly relevant and useful. The agreement discussed at this meeting is a new and very promising tool for mutually advantageous R&D ventures. At the same time, bilateral collaboration between the United States and single EU member states continues to play a significant and complementary role, especially for collaboration in those fields that are not included in the EU Framework Programme.

group also focused on possible applications of NGI technology: laboratory collaboration, meta-computing, and development of test beds. The group believed that specifying measures of success ahead of time would be important to the success of U.S.-EU collaboration on NGI. Measures of success mentioned included participation of individual researchers; a "high-technology" impact, that is, technological breakthroughs from U.S.-EU collaboration; and creation of economic value.

The NGI group also identified the following areas as possibilities for R&D collaboration: network dependability, network security, wireless technologies, portability, quality of service, scalability, middleware, and social impacts of e-commerce. Finally, the NGI group's next steps were similar to those of the other two groups: to define more precisely the process for collaboration and to choose specific projects for collaboration.

TRANSPORTATION CHALLENGES FOR THE 21ST CENTURY

John C. Horsely, Department of Transportation

Before beginning his summary of the transportation breakout sessions, Mr. Horsely, on behalf of Transportation Secretary Rodney Slater and Deputy Secretary Mortimer Downey, thanked the Board on Science, Technology, and Economic Policy and the Academy for convening the conference. He also thanked Wilhelmus Blonk of Directorate General VII of the European Commission for attending the discussion during the breakout sessions.

The breakout sessions covered three areas: intermodal transportation, intelligent transportation systems, and strategic enabling research. In general, Mr. Horsely stated that transportation is an important research topic on both sides of the Atlantic because it is such a pervasive part of economic and social life. An important goal of transportation research is to promote sustainable and competitive growth. Sustainable growth is an important concept for many economic activities in today's world, and transportation research must be brought into the concept of sustainable development.

Turning more directly to the breakout discussions, Mr. Horsely said that the need to "decouple the growth of traffic from the growth of the economy" was a pervasive theme emerging from the transportation breakout sessions. Traffic is growing rapidly in the United States and Europe, and eventually traffic congestion will inhibit economic growth. The transportation research agenda could therefore be very productively turned toward traffic congestion. In general, this means the application of information and other advanced technologies to transportation problems. Mr. Horsely reported that the breakout sessions also suggested examination of institutional barriers to implementing more efficient transportation systems in the United States and Europe.

Intelligent Transportation Systems

With respect to surface transportation, an important goal of transportation research over the past 10 years, which participants said should continue, is to inject information and communications technologies into all facets of the transportation system—roads, cars, buses, and rail. An example is the electronic tollbooth, by which a sensor at a tollbooth reads an electronic debit card installed on the car. The ultimate goal is to increase traffic flow, or throughput. In the United States it is estimated that throughput and associated efficiencies could increase by 15 to 25 percent through the use of intelligent transportation systems.

The group discussed some of the technical issues involved with intelligent transportation. These include network architectures, standardization, and interoperability. Further research would have to be done on individual applications, Mr. Horsely continued, including rail and freight transit and the human factors involved with intelligent transportation.

Maritime safety is another fruitful area for intelligent transportation systems, Mr. Horsely said. Applications in maritime safety include automatic ship-to-ship identification. Using the Global Positioning System, ships could instantly communicate speed, position, and conditions to other ships and to officials on shore. Such systems could help avoid collisions and spills of hazardous cargo, such as oil. As in other intelligent transportation areas, development of standards and common architectures remains a challenge to implementation.

Strategic Enabling Research

The discussion also touched on areas where common work between the United States and the European Union is possible. Areas in which "mutual exploitation" seems promising include logistics; monitoring and data collection; developing tools for forecasting demand; and human factors, such as training, workplace issues, and machine-human interfaces. Areas in which "mutual exploration" seems worthwhile are intelligent logistic systems, sustainability and air quality, operator fatigue, and advanced materials.

Intermodalism

Mr. Horsely said that the U.S. Congress has recently urged the Department of Transportation to explore further intermodal issues. With the goal of improved productivity increasingly driving business today, companies are looking for ways to shorten the supply chain. The supply chain was once two weeks long, Mr. Horsely noted, but that has now been shortened to two days or less in many businesses. Using transportation wisely, and choosing the right modes, could contribute to easing congestion and getting products to customers quickly. Transatlantic cooperation could explore the intermodal challenges in urban areas, including how to apply information technologies to logistical issues across transportation modes.

In summarizing intermodal challenges, Mr. Horsely reiterated the shared goal of decoupling growth in traffic from growth of the economy. As a next step he said that the transportation breakout groups recommend convening a workshop on institutional impediments to promoting intermodal efficiencies. Such a workshop could encourage participants to develop ways to use the right equipment to ship the right commodities to the right destination on time.

CLIMATE PREDICTION, FORECASTING APPLICATIONS, AND IMPACTS

John Krebs, National Environmental Research Council

Dr. Krebs opened his summary by saying that the climate group decided to focus its discussion on climate research. This area is important for countries worldwide: climate research is changing rapidly as more data are collected and analyzed and as our understanding of climate advances. Dr. Krebs noted in particular how improved computing power is improving researchers' ability to simulate climate.

In sounding a cautionary note on U.S.–EU collaboration, Dr. Krebs stated that the international climate research community is already very well connected. There are a number of umbrella programs to coordinate research, although not to provide research funding. Such programs include the International Council of Scientific Unions, the International Geosphere/Biosphere Program, the World Climate Research Program, and the International Human Dimensions Program. Given the existence of these programs, any collaboration between the United States and the European Union should coordinate closely with the broader international initiatives.

Research Priorities

The climate group's first topic of discussion was U.S. priorities for climate research. Bob Corell of the National Science Foundation presented a summary of a report from the National Research Council called "Overview of Climate Change: Research Pathways for the Next Decade." The group then heard a summary of the European Union's Fifth Framework Programme on climate change. The good news, said Dr. Krebs, is that the priorities reflected in these documents are very similar. This is not surprising because the international research agenda on climate changes is well understood and coordinated.

In the context of existing international ties and coordination, any new climate change research mechanism must be "more convenient than existing mechanisms if it is to work." For example, what would be unlikely to work is any mechanism that involves "double jeopardy," that is, one in which researchers must first clear hurdles in the United States and then clear similar hurdles in Brussels to obtain funding. Existing channels for international funding are available, with fewer burdens in the application process.

With the review of existing programs in hand, Dr. Krebs discussed the five areas in which joint U.S.-EU activity could add value to climate research.

Terrestrial Environment

The Kyoto Protocol requires that all signatory countries develop a carbon budget to understand where carbon comes from and where it goes. Understanding the terrestrial environment and carbon sink (i.e., how vegetation assimilates carbon) is important to developing carbon budgets. There are parallel and separate research programs in the United States and Europe. The climate group, reported Dr. Krebs, suggested that U.S. and EU programs be linked.

Predictability

Dr. Krebs pointed out the limits to predicting changes in climate that currently exist. In the short term it is accepted that forecasts are not reliable beyond 10 days. In the long term, climate researchers are confident about broad climate prediction in 50- or 100-year intervals. The fundamental theoretical question is: What are the limits of predictability on short- and long-term climate forecasts? That is, can we become more confident about short-term forecasts beyond 10 days? Can we predict broad climatic changes inside of 50 years? Another question raised by the group is: How exact must climate predictions be? Research should explore the gains that result from more exact predictions (in terms of benefits from actions to mitigate climate calamities versus costs of developing more exact models). All of these research questions can benefit from joint U.S.-EU research.

Comparison of Model and Impact Forecasts

The United States and several European countries (i.e., the United Kingdom, Germany, and France) have sophisticated models, and these models' performance should be compared. Currently, such comparisons do not occur.

U.S. National Impact Assessment Program

The United States has developed an assessment program to explore how climate change may affect specific U.S. regions. A number of European countries have national assessment efforts but have not broken down the assessments to regional levels. Acknowledging that the United States may be ahead in this area, the group suggested that the EU could learn from U.S. regional modeling efforts.

North Atlantic Oscillation

In Europe the North Atlantic greatly affects climate, in addition to changes in the mean temperature of the earth. Decade-to-decade changes in European climate appear to be affected by changes in atmospheric pressure between Iceland and the Mediterranean. More research is needed to understand the cause of the oscillation phenomenon in the North Atlantic. The issue is important to the climate of Europe and North America.

In closing Dr. Krebs mentioned two issues the group was unable to discuss but that are important nonetheless: (1) how to obtain data on areas of the world with small scientific communities (climate change is a truly global issue, and the United States and the European Union, with about 75 percent of the world's research and development spending, should pay attention to other areas of the world that do not have the resources for such research) and (2) research should consider the link between technologies that may provide a solution for climate change and the science of climate change itself. Looking toward the future, Dr. Krebs said that the United States and the European Union should further develop the list of areas in which collaboration could be fruitful. He noted that the climate group did not include representatives from industry or polar science (i.e., how climate change would affect the polar caps). Dr. Krebs concluded that broadening the discussion to include these constituencies would aid in refining the list.

HUMAN ENVIRONMENTAL HEALTH SCIENCES: ENDOCRINE DISRUPTORS

Paul Foster, Chemical Industry Institute of Toxicology

Dr. Foster began his summary by saying that the group's discussion focused on endocrine disruptors, that some consider to be a growing human health risk.[1] To briefly describe the endocrine disruptors issue, Dr. Foster said that there is a genuine and growing concern that chemicals released into the environment are having a serious impact on humans, wildlife, and vegetation. That is, chemicals are acting like hormones or affecting how hormones work.

Dr. Foster said that there has been a great deal of research on endocrine disruptors in the past three years, so the breakout group was not starting from ground zero. His group wants to build on existing work and try to build bridges

---

[1] Views on this topic and especially the assessment of the risk vary a great deal. The National Research Council has undertaken a study on this topic, entitled *Hormonally Active Agents in the Environment*.

across the Atlantic on endocrine disruptor research. As a first step, Dr. Foster said the group identified the following key issues in which collaboration may be fruitful:

- *Building a common language.* Despite the recent work on endocrine disruptors, there is no universal agreement on what an endocrine disruptor is. There are different definitions within EU member states and in the United States. U.S.-EU collaboration could help build a common language for endocrine dis-ruptors research.
- *The biology of endocrine disruptors.* The scientific community must deepen its understanding of biology to better understand endocrine disruptors. We do not fully understand what is normal in living organisms, so it is difficult to determine what is abnormal. This issue is especially important with respect to reproduction and early childhood development.
- *Improve understanding of the impact of hormones on disease.* Scientists must make more progress in understanding how hormones or hormone-like agents affect disease. If we better understand this, scientists can test and screen for chemical agents. As the U.S. Environmental Protection Agency (EPA) lists over 70,000 chemicals, it is important to know which chemicals to test. This area would be a prime candidate for collaboration because common protocols are needed, testing methodologies must be validated, harmonization must occur, and international trials must be coordinated. Also, to ensure that comparisons of different test results in a region are valid, scientists must use the same testing methods in the same part of the world.
- *Risk assessment.* The United States and the European Union would benefit from collaboration on risk assessment—for example, determining the frequency of occurrence of compound X and the magnitude of its harm.
- *Classification and labeling.* This is a potentially thorny issue as it requires agreement on whether risk is communicated to citizens or some notion of intrinsic hazard. Most labeling in Europe is hazard based, whereas the scientific perspective would have a preference for risk-based labeling.
- *Exposed populations.* If we know what the effects of a chemical are and the risks associated with exposure, it is equally important to know the extent of exposure among key populations (i.e., humans, wildlife, vegetation). Knowing the exposed population, along with affects of chemicals, could allow scientists to make cause-and-effect claims about the presence of chemicals among certain populations.
- *Exposure assessment.* As noted above, the EPA lists over 70,000 chemicals, and the general population is not exposed to all equally. U.S.-EU collaboration could be particularly helpful in determining the level and frequency of exposure to certain chemicals.

- *Technologies for remediation.* If it is established that a certain chemical creates risk, what should be done? Banning a chemical may be problematic given the pervasiveness of certain chemicals in society. There is already a large body of research in this area, and U.S.-EU collaboration could develop a global inventory of available research on technologies for remediation. Scientists must do more than just communicate the content of their research; they must also provide information such as the size and scope of their research so as to bring the right expertise together worldwide. Dr. Foster noted one model of how to evaluate data from such studies, namely, the U.S. National Institute for Environmental Health Sciences. This institute brings together experts to evaluate studies on reproductive health.

Dr. Foster concluded his presentation with the following observations:

- *Funding mechanisms.* The research community must know the procedures for obtaining funds. As noted in another breakout session, scientists will not seek funds from a source if the procedures are too difficult.
- *Collaboration.* A panel of experts could usefully be convened to fully describe the opportunities for collaboration and to be responsive to the Fifth Framework Programme that gets under way in 1999. The breakout session only touched on possible areas for collaboration, and specific proposals must be developed to realize collaboration.
- *Trade issues.* One breakout participant noted that endocrine disruptors involve chemical management, and when governments try to manage chemicals they have the potential to affect trade.

Opening Remarks

John Cadogan
Director General, Research Councils of the United Kingdom,
for the U.K. Presidency of the European Union

Dr. Cadogan began his address with the observation that the large and distinguished group of scientists, engineers, and industrialists in attendance highlighted the interest in scientific collaboration on both sides of the Atlantic. Although this underscores the economic importance of science and technology, Dr. Cadogan implored conference participants not to forget curiosity-driven science. He noted that even though directed research programs can create great wealth and prosperity, the role of the individual pursuing his or her research for the sake of advancing knowledge must not be overlooked. There must be unfailing support for directed research, but it is also important to shine light on curiosity-driven inquiry.

No committee, no government, no board of directors, and no civil servant ever made a discovery, Dr. Cadogan stated, let alone a development. Discoveries can only be made in a laboratory. Everything that the United States and the European Union may do with respect to collaboration will come to nothing unless the creativity of scientists and researchers is released in the laboratory. He observed that we all may sometimes mistakenly conclude that events—such as conferences or appearances on TV—are more important than results. We must not lose sight of the process of discovery and the role of the individual researcher in driving discovery.

Many of the most important discoveries, Dr. Cadogan continued, occurred when scientists set out with one purpose and wound up discovering something very different. It is too easy to say: "That won't work, obviously," only to find out later that the "impossible" experiment yielded remarkable findings. Some of the landmark breakthroughs that came about in this fashion include antibiotics, the laser, nuclear fission, the discovery of DNA, the ozone hole, and semiconductors. All of these were developed when people were not really looking. None of these discoveries were predicted, and many were discounted immediately upon the discoveries being made known.

The nature of these discoveries reminds us of two important lessons of history, Dr. Cadogan continued. First, there are no limits on the advance of scientific knowledge, and, second, we often forget lesson one. Many of our most eminent inventors and scientists have fallen victim to lesson two. We are not driven enough to think the unthinkable. Even many of our best innovators lose their vision after they have accomplished a great deal. Dr. Cadogan shared some examples of this phenomenon:

- Alexander Graham Bell, shortly after he invented the telephone, predicted that one day *every* manufacturing firm in the United States would have a telephone.
- The chief engineer of the British Post Office said in 1876 that the telephone might be all well and good for the United States but that it would never catch on in Great Britain because the country has an adequate supply of messenger boys.
- Ten years later the same chief engineer of the British Post Office said that if the growth of telephone subscribership continued, by the year 2000 every woman in Great Britain would have to be a telephone operator.

Dr. Cadogan thus cautioned against scientists and industrialists thinking that nothing more is discoverable. Society must nourish creators *and* innovators. These are rarely the same people, and they can be difficult to work with. Moreover, they are unlikely to welcome advice from governments, politicians, or civil servants. But we must nonetheless cultivate the creators, who dream of new things, and the innovators, who make the new things work in the marketplace.

Dr. Cadogan added that Europe has a great deal to learn from the United States in the business of innovation. He also observed that, while Europe had grown quite skilled in collaborative research, and while scientific inquiry remained vibrant in Europe, Europe could do better at "cracking the tough ones" in some research areas.

Turning to the impact of discoveries, Dr. Cadogan noted that most scientific advance was incremental. A scientist must often be content with "putting a brick in the wall" and being satisfied with the entire edifice, once it is built through the efforts of many scientists. Only a few of us are given the ability to make the startling breakthrough that changes the world. In fact, there are really only four or five discoveries that have changed the world in this century—the understanding of organic and physical chemistry at the start of this century, which led to the chemical industry; manned flight; nuclear fission; the transistor; and the genome. Each of these discoveries or developments has had widespread impacts on the world, and most will continue to alter the shape of society. Dr. Cadogan said that stunning discoveries, some of which we cannot even conceive of today, will be made on the basis of the genome. The innovators will have a "wonderful time" with discoveries coming from genome research.

Turning to the program's next speaker, Gordon Moore, Dr. Cadogan commented that Gordon Moore knew more about innovation than "I've had hot dinners." Noting Dr. Moore's standing as one of Silicon Valley's founding fathers, Dr. Cadogan observed that Europe has long marveled over the creativity and economic vitality of Silicon Valley. Perhaps, Dr. Cadogan concluded, Dr. Moore could tell us in his remarks where "Genome Valley" will be.

Keynote Address

Gordon Moore
Chairman Emeritus,
Intel Corporation

Dr. Moore opened his address by complimenting conference participants on the high level of discussion he had heard that day in terms of both technical sophistication and the ways in which scientific advances could benefit society. In previewing his remarks Dr. Moore said that he would have a different emphasis, focusing on the practical issues facing the research community and the semiconductor industry. From this perspective four areas stand out as important: the changing environment for research and innovation, the evolution of the semiconductor industry, future challenges for the industry, and the role of international cooperation in meeting future challenges. Noting that his career had focused on the details, where he had tried to build new and strong structures on the foundations of others, Dr. Moore hoped that his remarks would shine light on how to build foundations for the future.

CHANGES IN THE RESEARCH AND DEVELOPMENT ENVIRONMENT

The contributions of large industrial research laboratories, which have been so important in the past, have been diminishing in recent years. Competitive pressures and corporate downsizing have prompted the reduction in the size of industrial research laboratories. Corporate research and development (R&D) has also become much more short term in the past several years. It is harder than ever for corporations to capture the fruit of an unexpected R&D breakthrough. Thus, they tend to focus on R&D that is closely related to their core businesses and that is "reasonably predictable."

Even with the phenomenal contribution that research has made to society in the past several decades, it would be difficult to make a case that the companies that conducted the research captured its fruits. This makes it difficult for corporate leadership to justify fundamental long-term R&D. It is more sensible to focus on the short term, on research that is directly related to the business.

However, the need for fundamental research is as urgent as ever—maybe more so than ever before. But because of competitive pressures on corporations, we must look elsewhere for such research. That is where universities and government-supported laboratories come into the picture. Dr. Moore said that he had a bias toward university research when the issue is framed as a choice between directing funds to universities or to government laboratories. As he explained, with university research, "even if the research fails, you still get the students." Increasingly, however, the output of university research, not just the supply of trained researchers, is important.

EVOLUTION OF THE SEMICONDUCTOR INDUSTRY

Turning to the semiconductor industry, Dr. Moore noted how the industry had been a direct beneficiary of the industrial research system. The invention of the transistor at Bell Laboratories was a seminal event, signaling the birth of an industry. To convey a sense of the industry's size, Dr. Moore pointed out that the transistor is the highest-volume manufactured product in the world. Each year there are more transistors manufactured than printed characters of any type—newspapers, magazines, books, and copies of documents. There are about as many transistors made each year—10^{16} to 10^{17}—as there are ants on the entire planet. Put still another way, each year the semiconductor industry makes 10 million to 20 million transistors for every person on earth.

Declining Unit Costs

In addition to the size of the industry, Dr. Moore said that the other distinctive feature of the semiconductor industry was declining costs. Several years after the invention of the transistor, a study estimated that the cost of making a single transistor would soon fall to 60 cents. Today, a 64-megabyte dynamic random access memory (DRAM) chip with approximately 65 million transistors costs $8. That is a little less than one-eighth of a microdollar per transistor, or 120 nanobucks. The continuous decrease in costs and the corresponding improvement in performance have driven semiconductor industry revenues to $150 billion annually, and the industry in turn supports a $1 trillion electronics industry.

Manufacturing Challenges

Commenting on manufacturing challenges facing the industry, Dr. Moore remarked that the semiconductor industry stands out among high-technology industries for its complexity and fast-changing technologies. Some semiconductor product lines, such as DRAMs and other memory chips, have become commodities. Others, such as higher-end microprocessors, have a great deal of intellectual property embedded in them and command substantial margins in the marketplace.

Semiconductor technology is also very flexible in that there are many market niches and specialty products. The technology enables custom products to be made for specific applications or firms. As a result of the technology's flexibility, there are hundreds of semiconductor companies and thousands of firms that support them.

Rapid Technological Change

Rapid technological change is another defining characteristic of the semiconductor industry, Dr. Moore continued. As an example of fast product obsolescence, 80 percent of Intel's revenue in 1997 came from products that had not been introduced as of January 1 of that year. New products mean new plants. The plants themselves today cost multiple billions of dollars, and the equipment within must be replaced every few years. After several generations of equipment, semiconductor plants must be scrapped and new ones built. The only constants are extremely rapid change and falling prices; on average, prices for semiconductor devices fall 20 to 30 percent per year, with the price of some devices falling as much as 50 percent in a single year.

There is no other industry like the semiconductor industry, and it leaves much room for blunder by corporate leadership. Dr. Moore noted that, for this reason, industry leadership has changed many times in the years of his involvement in semiconductors. He recalled that he helped found Fairchild Semiconductor in 1957, which was an industry leader into the mid-1960s. Today, even though the Fairchild name has been revived, the company no longer exists as a semiconductor maker.

The industry's dynamism creates enormous opportunity for start-up companies, Dr. Moore continued. The strength of U.S. high-technology industry relative to that of other nations has been its ability to generate start-up companies to exploit technological opportunities. From his own experience, Dr. Moore said that it was much easier to move swiftly and efficiently in developing new things in a start-up environment. A small focused group of committed entrepreneurs can accomplish tremendous things. As a company grows, it inevitably develops a bureaucracy that makes it difficult to develop new products or exploit new technologies. Dr. Moore also emphasized that start-ups rarely develop new technologies but typically find new ways to exploit them in the market. Silicon Valley contains numerous examples of this.

Smaller Is Better

With the declining price of the transistor, Dr. Moore said that the industry has been operating under a variant of Murphy's law: "As things get smaller, everything gets better simultaneously." As the industry packs more circuits on a chip, performance improves as circuit connections grow shorter, thus increasing operating speed. Moreover, with more of the system residing on a single chip, the entire system becomes much more reliable. All this occurs, Dr. Moore reiterated, in an environment of dramatically falling costs.

Silicon Real Estate

Recasting the issue somewhat, Dr. Moore said that what the semiconductor producer sells is real estate—the area of the 200-mm (by today's standard) silicon wafer. On average, that silicon wafer represents $1 billion per acre, maybe a half a billion for DRAMs and several billion for microprocessors. The main challenge for the industry is really a real estate development problem: how to pack more circuits on the wafer with greater functionality in such a way that, while maintaining historical rates of price decreases, the revenue stream from the wafer remains at $1 billion per acre.

FUTURE CHALLENGES FOR THE SEMICONDUCTOR INDUSTRY

Dr. Moore began his discussion of future challenges with some historical observations on the industry's marketplace evolution. In the past the semiconductor industry grew by "assimilating the value added of its customers." By that Dr. Moore meant that innovations such as the integrated circuit allowed, indeed required, semiconductor companies to take on the design tasks that its customers—such as computer systems makers—had done. This was a tough sell, essentially telling a customer that may have had a division specializing in circuit design that its division was no longer necessary. This led to Bob Noyce's second great contribution to the semiconductor industry, namely, the idea that semiconductor firms should sell integrated circuits to system houses for less than it would cost them to assemble the circuits themselves. Noyce's idea led to the growth of high-volume manufacturing in the semiconductor industry and to the cycle of declining costs.

This development fundamentally changed the industry's interface with its customers. Semiconductor firms began engaging in circuit design, then moved into logic, and today new computer designs are essentially done by the semiconductor industry.

Approaching Limits

As for the future, Dr. Moore said that it was his belief that the current technology for semiconductor manufacturing, photolithography, would be able to provide advances in performance and declines in cost for another three or four generations. At that point the atomic nature of matter would begin to be an impediment to further advances from photolithography. The wavelength of light used in photolithography would become larger than the desired circuit features to be etched onto the silicon wafer.

The Roadmap

Before we approach those limits, Dr. Moore stated that a great deal of research is necessary to get the most out of photolithography. The main guide for this research agenda is the National Technology Roadmap for Semiconductors, developed by the Semiconductor Industry Association and managed by SEMATECH. The roadmap, by extrapolating technology needs into the future based on past performance, lays the track so that the semiconductor industry's technology locomotive can keep moving forward. The roadmap has proved to be a powerful concept in the semiconductor industry, and Dr. Moore said that it is a tool that other industries could emulate. Its main virtue is defining the research agenda for the mainstream of an industry and charting a way to meet goals.

Beyond Photolithography

Moving beyond photolithography will present new challenges. Dr. Moore described three techniques to replace photolithography, which have varying degrees of promise:

- *X-ray lithography*—the use of x-ray shadow masks has been tried; however, there are still substantial problems with this approach.
- *Electron optics*—this area has shown some promising early results, but there is still a long way to go from the laboratory to production.
- *Extreme ultraviolet (EUV)*—using normal optical techniques in conjunction with multiple reflectors holds promise, yet EUV may not be able to use light that is smaller than 13 nm in wavelength.

Two further challenges affecting the industry's future are *design*—in a few years, a chip will contain 1 billion transistors, and it will be a huge challenge to improve design techniques so that ever-more complex designs can function reliably, and *change in wafer size*—Dr. Moore noted that the conference would hear more about the change to the larger (300-mm) wafer size when Bill Spencer, in his presentation, described the I300I initiative of SEMATECH.

Dr. Moore explained that, in the past, equipment and plant costs were low enough that a single company could change wafer size on its own. But today equipment costs are so high that, as the industry shifts from 200-mm wafers to

300-mm wafers within a few years, all must change at once. The equipment manufacturers simply cannot afford to simultaneously manufacture two generations of equipment.

ROLE OF INTERNATIONAL COLLABORATION

With respect to the role of international collaboration in the research challenges facing the semiconductor industry, Dr. Moore remarked that he is unsure whether government-to-government collaboration is the correct approach. However, he said that international collaboration among firms and universities holds great potential. Because the research challenges in making devices with circuits with ever-smaller line widths are enormous, international collaboration might be useful, as it is in the conversion to the 300-mm wafer size. The change to 300-mm wafers is, of course, already a subject of international collaboration in the I300I project. Some radically new approaches to semiconductor and computer technology, such as quantum dots, quantum computing, and DNA computing, are possible candidates for international collaboration, particularly among universities.

In conclusion Dr. Moore said that the semiconductor industry has had a phenomenal run in exploiting new technologies. The industry has benefited from the support of the U.S. government and the industry's own initiatives, such as the Semiconductor Research Corporation, which develops new technologies at universities, and SEMATECH, which focuses on manufacturing technology. However, large research challenges must be met if the industry is to continue its advance. Collaboration is clearly one means of meeting these challenges. Dr. Moore closed by saying that he was gratified to see an entire conference devoted to scientific and technological collaboration. He expressed the hope that expanded transatlantic collaboration would result in a breakthrough with as big an impact on society as the semiconductor.

Second Day's Welcome

William Wulf
President
National Academy of Engineering

On behalf of the Academy let me welcome you all to the second day of our meeting on transatlantic science and technology (S&T) cooperation. Yesterday, the morning session addressed trends in transatlantic S&T policy and international cooperation. That was followed by some fascinating breakout sessions on how our two communities, with their proud traditions, can collaborate in cutting-edge technologies, namely information technologies, transportation, climate change, and health sciences.

Today, we intend to concentrate on some of the public policy issues that may arise as collaboration between the United States and the European Union (EU) unfolds. We plan to do this by looking at "best practice" in several S&T cooperative efforts and ask some questions: What do we know about past and current efforts in domestic and international S&T cooperation? What lessons from these efforts should we take into the future of S&T cooperation between the United States and the European Union? As our agenda indicates, we plan to address best practices by looking at three engines of economic growth associated with all businesses but with a focus on high-technology:

- *Small business.* A key element of economic growth is a vibrant high-tech small-business sector that creates a self-sustaining cycle of job creation and growth. The United States and Europe have programs to foster small-business growth, and our first session this morning explores several of these programs.
- *International research and development.* R&D is becoming more costly and funding for long-term R&D is increasingly scarce. In this environment it is natural for countries and companies to look for partners around

the globe. We are privileged to have with us today Bill Spencer, of SEMATECH, to discuss a key international R&D effort in the semiconductor industry.
- *The technical work force.* Just as more and more companies search the globe for R&D partners, they also look worldwide for technical talent. This creates both new opportunities for transatlantic cooperation and challenges in managing R&D efforts with an international work force. Our third panel this morning, which I will moderate, will look at these issues.

The U.S.-EU S&T agreement is a valuable opportunity for both sides. It is too valuable to let issues of management and implementation fend for themselves. We can learn much from each other, but we must also be aware of common policy challenges inherent in cooperation. Today, and in the follow-up conference in Europe planned for 1999, we hope to shed some light on how to meet those policy challenges. This will give a new era of U.S.-EU cooperation operational life. Cooperation, like marriage, requires constant dialogue to ensure that it is productive and sustainable. By beginning a dialogue today and recognizing that it must continue, we dramatically increase the chances of the agreement bearing fruit for the United States and the European Community.

Best Practices in Small-Business Technology Development Programs

Helmut List
Chairman
Industrial Research and Development Advisory Council, Austria

Dr. List opened the session by observing that technology development for small businesses is an important topic for both the United States and Europe. Deploying technology in small businesses is a highly effective way to enhance innovation, create jobs, and spur business growth. The Industrial Research and Development Advisory Council has been promoting ways to make the Fifth Framework Programme more accessible to small and medium enterprises (SMEs) in Europe.

Recalling Ambassador Hugo Paemen's earlier comment that what fits for small business development on one side of the Atlantic may not fit on the other, Dr. List expressed the hope that this session would allow both sides to learn from one another. He stressed, however, that we should do more than compare notes on our respective experiences. We should begin the process of making transatlantic cooperation work for SMEs on both sides of the Atlantic.

In the area of high-technology start-up companies, Dr. List said that the United States is the model for the world. Europe is trying to create an environment in which small-business start-ups can thrive, and it hopes to draw lessons from the United States. A key feature of the U.S. environment is the availability of venture capital. Having capital available early in the innovation process is critical for SMEs. Although Europe has lagged the United States in developing a vibrant venture capital industry, Dr. List believes it is now making strides in this area.

In broader terms, Dr. List said that there are two reasons for the growing importance of small businesses in the economy:

- The original equipment manufacturing (OEM) industry has been restruc-

tured. OEMs are turning to smaller companies, with entire systems or subsystems being contracted out to small firms.
- Technology and the shortening of product cycles have caused rapid changes in the market. These changes create opportunities for small firms, which can move more quickly than larger ones.

Dr. List concluded that Europe and the United States must create stronger links between small businesses, universities, and government labs. Such a multidisciplinary approach could truly open new doors for small businesses.

INDUSTRY-LABORATORY COOPERATION: THE AMTEX EXPERIMENT

Jerry Cogan
Millikan Research

Dr. Cogan began his comments by describing a conference that the national laboratories held in 1992 that brought together industries that the labs had not dealt with in the past. The textile industry participated in the conference, although many people in the textile industry and the national labs wondered whether they could develop a meaningful relationship with the labs. The textile industry was largely unaware of the specific capabilities of the labs, other than knowing that U.S. national laboratories held a tremendous store of technology. At the conference lab officials asked representatives from the textile industry to consider ways in which the textile industry and labs could become partners.

Issues in Industry-Laboratory Partnerships

A key question that laboratory officials posed to representatives from the textile industry was: How important is your industry to the overall economy? Dr. Cogan presented figures showing the size of the textile industry relative to the entire manufacturing sector. Although a very disaggregated industry, with approximately 25,000 separate companies, the textile industry in 1992 had 9 percent, or 1.7 million, of all manufacturing jobs in the United States. The industry was, however, losing jobs and market share; from 1992 to 1996 the number of jobs in the textile industry fell from 1.7 million to 1.2 million. The industry was also losing market share to foreign producers.

A second question raised was whether the U.S. textile industry could make effective use of laboratory technology. The answer was strongly affirmative. Dr. Cogan observed that heavy use of technology in the textile industry would come as no surprise to Europeans, because the European textile industry is very technology intensive. From the U.S. perspective, when an industry is struggling to survive, as textiles have been, modernization is the strategy to undertake. The textile industry's efforts to modernize are reflected in investment data; the industry's capital investment is higher than that of other manufacturing industries and has been, on average, for the past 10 years.

In short, the textile industry and the laboratories quickly realized, said Dr. Cogan, that lab-textile industry partnerships could be a "real win-win." Because the textile industry is so diffuse—25,000 separate firms—it was necessary to find a way to organize a lab-industry partnership in a way to involve the entire industry. Fortunately, the industry was already organized for research when the lab partnership opportunity presented itself, with collaborative efforts under way at a number of universities. The industry had also organized a National Textile Center to conduct research.

As a result of these initial contacts, the American Textile (AMTEX) Partnership was launched through a cooperative research and development agreement (CRADA) with the national laboratories. The CRADA covered a number of projects at the following national laboratories: Argonne, Brookhaven, Idaho, Lawrence Berkeley, Lawrence Livermore, Los Alamos, Oak Ridge, Pacific Northwest, Princeton, and Sandia.

The Demand-Activated Manufacturing Architecture Project: Reducing Waste

As an example of one lab-industry project under AMTEX, Dr. Cogan described the Demand-Activated Manufacturing Architecture (DAMA) project. DAMA is important to industry because it addresses waste. The waste of materials used in textile production represents a $45 billion problem annually for the industry; this amounts to 25 percent of U.S. retail apparel sales. DAMA's objective is to reduce waste by using tools of electronic commerce to better manage inventory and advanced software and computers to reduce the time to market.

In concluding Dr. Cogan listed several elements of the lab–industry partnership in textiles that he suggested may serve as lessons for other partnerships: involve all of the industry; have industry develop technology roadmaps; focus on precompetitive technologies (i.e., manufacturing process technologies and systems); make sure that the partnership contributes to the core missions of the public partner; define the benefits for partners at the outset, including intellectual property rights; and capitalize on existing government resources so that government does not expand existing staff or facilities.

LABORATORY PARTNERSHIPS WITH INDUSTRY

Dan Hartley
Sandia National Laboratories

Dr. Hartley said that his remarks would focus on why the national laboratories seek partnerships with industry, what gives the labs the legal right to do so, and why the labs think it is important to have close ties with industry. Although the AMTEX partnership is important, he wished to address these broader issues.

Increasing Cooperation with Industry

Dr. Hartley characterized his job at Sandia as one of looking to the future of the lab, which in effect means increasing collaboration with industry. Looking at Sandia's mission historically, he noted that for 50 years its major mission has been to develop nuclear weapons for the Department of Defense (DOD). Starting in the 1970s, Sandia began to undertake other weapons missions for DOD; in the 1980s it turned to energy-related work for the U.S. Department of Energy (DOE). More recently it has focused on critical infrastructure issues and worked with other countries on handling nuclear materials. From the mid-1980s to the present, Sandia has been working more closely with private industry.

Sandia's Core Mission

In describing Sandia's traditional mission more fully, Dr. Hartley said that a typical nuclear missile contains about 6,000 parts. Sandia makes approximately 5,500 of those parts—essentially everything but the nuclear components of a weapon. Livermore and Los Alamos labs make the nuclear devices for weaponry. Sandia's role in making over 5,000 parts gives the lab the same expertise as any large manufacturing firm, whether it is in consumer electronics or auto manufacturing.

The Push Toward Computer Simulation

Changes in the law have led Sandia to rely more heavily on computer simulation in recent years. At one time the U.S. government could test a weapon by underground explosions in Nevada. That is now prohibited, so Sandia must turn to computer simulation to test weapons. This is known as Science-Based Stockpile Stewardship and is a program managed by DOE.

Sandia has also expanded into antiterrorism because the nature of worldwide threats has changed. Sandia makes a highly sensitive explosives detector that is used in airports. Another new component to the Sandia mission is in the disposition of nuclear waste. Dr. Hartley said that Sandia designed a salt cavern in southern New Mexico in which low-level nuclear waste is stored.

Infrastructure

With respect to critical infrastructure, Sandia explores the survivability of major systems. As an example Dr. Hartley described Sandia's role in addressing

Japan's concern about the construction of a nuclear power plant near a U.S. air base in Japan. The Japanese government was worried about the safety implications of a potential crash of a U.S. plane into the reactor of the power plant. Using computer simulation technology, Sandia was able to assure Japan that a plane could not penetrate the reactor's wall and that a plane crashing into the reactor would be reduced "to powder" rather than cause a nuclear incident.

Industry's Role

Turning to industry's role in Sandia's mission, Dr. Hartley emphasized that it was crucial that Sandia work with industry. Such a relationship helps the laboratory keep abreast of new scientific and technical developments, and the lab shares in the gains, and costs, of collaboration with industry. To capture the relationship, Dr. Hartley recalled a story regarding the Goodyear Tire Company when he was preparing congressional testimony. Dr. Hartley had asked Goodyear for a sentence or two to convey why it thought collaboration with Sandia was worthwhile. A Goodyear official responded by saying: "Goodyear may have a job that requires A + B. Sandia may have a job that requires A + C. Why not work together on A?" In sum, Dr. Hartley said that all of Sandia's missions can benefit from work with industry.

In addition to lab-industry collaboration aiding Sandia in remaining on the cutting edge technically, Dr. Hartley noted that changes in federal law in recent years have encouraged collaboration. The Bayh-Dole Act, the Stevenson-Wydler Act, and the Federal Technology Transfer Act have all encouraged the national labs to work closely with the private sector.

From the perspective of the private sector, Dr. Hartley identified two reasons why he believes industry turned to Sandia and other national labs: a concentration of world-class facilities and a businesslike culture, especially at Sandia. For 45 years Sandia was run by AT&T, and it has been run by Lockheed Martin for the past five years. This has given the lab an appreciation of how to cooperate with industry to carry out its mission.

Dr. Hartley then provided an outline of Sandia's world-class facilities:

- *Microelectronics.* Sandia's microelectronics facility houses a number of advanced projects, including one on microelectromechanical systems (MEMS). MEMS are tiny machines whose dimensions are as small as a red blood cell that can bring computing power to new applications. Intel is one of many industry partners working on MEMS at Sandia.
- *Manufacturing.* It takes world-class manufacturing capability to build the 5,000 parts that Sandia must make for its weapons mission. Sandia's manufacturing facility can serve as a model for industry. As one example Dr. Hartley pointed to Sandia's computer-aided design technology, which

enabled users to move from computer design to a three-dimensional structure in one step.
- *Robotics.* Sandia has what may be the most modern robotics facility in the world. One project involves small robots that serve as "collective agents" in the battlefield as they look for intruders. Each robot perceives only a portion of the environment, so the robots must communicate with one another to collectively locate the intruder. Dr. Hartley also said that Sandia is working with Lockheed Martin to develop an intelligent robot for painting aircraft that does not inadvertently punch through the aircraft body while painting.
- *Teraflop computer.* This is part of DOE's Advanced Strategic Computing Initiative, and Sandia's teraflop computer, built by Intel, is the fastest computer in the world. The computer simulated the plume of the Shoemaker comet before it struck Jupiter, and the simulation compared very well to the actual plume captured by the Hubbell telescope.

Collaborating with Small Business

In closing Dr. Hartley said that Sandia places great priority on engaging SMEs in the lab's activities. Sandia has logged over 1,300 small business "assists" and has over 300 CRADA partners. Sandia's CRADA partnerships are located in all regions of the country. Although the laboratory is open to increased cooperation, Dr. Hartley emphasized that Sandia does not engaged in "job shopping," that is, entering into partnerships just for the sake of doing so. Sandia believes in the value of partnerships but also that they must be done right and carefully specified ahead of time to be mutually beneficial.

THE U.S. EXPERIENCE WITH THE SMALL BUSINESS INNOVATION RESEARCH PROGRAM

Joshua Lerner
Harvard Business School

Dr. Lerner began his remarks by observing that the growth of high-technology clusters in the United States (e.g., Silicon Valley, Route 128 in Boston) has generated a great deal of interest abroad in how to emulate such technology-driven successes. There is a strong and sensible intuition that high-technology industries are sources of job and income growth, although the precise mechanisms through which to foster such growth remain subject to debate. A key question Dr. Lerner proposed to address in his presentation is the role that public venture capital programs have played in developing clusters of high-technology economic activity in the United States.

Dr. Lerner recalled some history of public venture capital efforts in the United

States, noting that U.S. public venture capital programs have historically been sizable. From 1958 to 1969 the Small Business Investment Corporation (SBIC) provided $3 billion to small firms, which was three times the sum provided by private venture capital firms. In 1995 small business financing programs provided $2.4 billion in funding, compared with $3.9 billion provided by private venture capital funds to small businesses. Beyond the size of public venture capital programs, they have been reputed to be important to the early success of well-known companies, such as Apple, Chiron, Compaq, FedEx, and Intel. Moreover, public venture capital programs have served as training grounds for private venture capitalists. Many leading figures in the U.S. venture capital industry were part of the SBIC program of the 1960s. Dr. Lerner noted that countries with vibrant venture capital sectors, such as Israel, Singapore, and Taiwan, each have public venture capital programs as complements.

The association between public venture capital programs and private venture capital does not establish causality; it may be a historical accident that public venture capital programs and the development of clusters of high-technology U.S. firms have coincided with one another. However, if it is not a historical accident, it is necessary to explore the mechanisms by which public venture capital programs are translated into innovative behavior in the economy.

To address these issues Dr. Lerner proposed to examine the Small Business Innovation Research (SBIR) program. The SBIR program was enacted in 1982 and requires that federal agencies with extramural R&D budgets in excess of $100 million set aside a portion of their budgets for awards to small business. Initially, 1.25 percent of the agencies' R&D budget was to be set aside for SBIR; this figure was increased to 2.5 percent in 1992. For fiscal year 1996 the 2.5 percent set-aside resulted in $1.1 billion in funding for the SBIR program.

Dr. Lerner described two theoretical motivations for government assistance to small business:

- *R&D spillovers.* In generating the know-how that underpins new technology, knowledge usually flows somewhat freely among a technical or scientific community. A firm investing in technology therefore cannot capture all of the knowledge that goes into creating a new product or process. In economists' parlance, such a positive externality is a social good but leads to underinvestment in knowledge by a private firm. Government assistance is thus justified to make up for the underinvestment.
- *Information asymmetries.* A high-technology entrepreneur will usually know a great deal more about a technology and its market potential than a banker who may be considering extending a loan to the entrepreneur. A venture capitalist can serve an intermediary function between banker and entrepreneur. In other words, the venture capitalist provides a signal to the banker by investing (or not) in a small high-technology start-up firm. Such certification by a venture capitalist is useful, but venture capitalists fund

only a small fraction of start-ups. Public venture capital programs can fill this gap and serve as an additional certification mechanism for private capital markets.

Even with the theoretical benefits of government assistance, Dr. Lerner raised the issue of potential problems in public venture capital programs. Distortions could exist if public venture capital programs favor certain entrepreneurs who may have well-established channels to policymakers. Officials in public venture capital programs may also give grants to firms destined for success, so that programs are judged positively when they are assessed.

Before discussing the results of his study, Dr. Lerner said that it is generally thought that the SBIR program functions well. The dispersed nature of the program is a virtue. SBIR is spread out among 11 agencies, and grants are usually no more than $750,000, small enough by federal standards to attract little attention. Agencies, moreover, have taken steps to keep political interference to a minimum in allocating awards. However, there have been concerns raised about the program, and they fall into two categories:

- *Regional distribution.* SBIR awards have gone predominantly to areas with a concentration of private venture capital, such as California and Massachusetts. Some believe that there should be a wider geographic distribution of awards.
- *Clustering of awards by institutions.* Some organizations win a large number of awards, suggesting that winning awards has become an end in itself, as opposed to commercializing new technology.

In his study Dr. Lerner examined the long-term impacts of the program, rather than relying solely on anecdotes for evaluation. He looked at 1,435 small firms over a 10-year period, with some firms being SBIR awardees and others being nonawardees with characteristics that were similar to the awardees. The SBIR program itself is very competitive, with about 5 percent of applicants winning grants; although firms with fewer than 500 employees are eligible, awardees tend to be much smaller than that. Dr. Lerner's analysis showed that:

- SBIR awardees experienced stronger job growth over time than nonawardees. Employment at SBIR awardees grew by 26 employees on average over the 10-year time horizon of his study versus job growth of 6 employees on average among nonawardees.
- SBIR awards went to regions with active venture capital sectors, such as California and Massachusetts.
- The first few SBIR awards have the strongest impact on job growth, with later awards showing little effect.

Overall, Dr. Lerner concluded that the SBIR program has had a positive impact on high-technology start-ups that have won SBIR awards. There is room, however, for fine-tuning implementation of the program. With the SBIR program scheduled for congressional reauthorization in 1999, Dr. Lerner said that there would be an opportunity to consider improvements.

Comments from the Audience

A participant observed that Dr. Lerner had talked about the private return to participation in SBIR by firms but not the government's return. Dr. Lerner was asked to comment on the government's return from the SBIR program. In response, he acknowledged that his focus on the private return is a limitation of his analysis. There is, however, great difficulty in assessing social returns because it is hard to separate the portion of investment activity that would have occurred in SBIR awardees from the portion that the program induced. Even if a good bit of the investment activity would have taken place without SBIR, Dr. Lerner said that the R&D spillovers generated by SBIR firms would be an additional benefit.

Dieter Seitzer, director of the Fraunhofer Institute, in Erlangen-Nürnberg, asked for examples of how SBIR awards work in agencies. Dr. Lerner commented on the tremendous diversity among agency approaches to implementation and SBIR awardees. With respect to implementation, agencies such as DOD and the National Aeronautics and Space Administration use SBIR as a procurement tool for technologies with very specific purposes. Other agencies, while adhering to SBIR's mandate to carry out agency missions, may fund technologies with longer time horizons before payoff. Regarding diversity, Dr. Lerner said that SBIR awards run a wide gamut of technologies, in contrast to venture capitalists, who tend to fund a narrower range of "hot" technologies, such as Internet technologies today.

A participant asked whether it was possible to compare the results of SBIR with those of venture capitalists. Although he had not done a systematic comparison, Dr. Lerner's strong suspicion was that firms funded by venture capitalists perform better than those funded by SBIR. Venture capital-funded firms undergo strict scrutiny, and only a small fraction of companies seeking venture capital receive it. It is important to recognize that the goals of SBIR differ from those of venture capitalists, given SBIR's focus on agency missions. In fact, Dr. Lerner added, there might be a great concern if SBIR focused only on biotechnology firms, as many venture capitalists do today, as opposed to the wide variety of firms that the SBIR funds.

THE EU EXPERIENCE WITH SMALL- AND MEDIUM-SIZED ENTERPRISE DEVELOPMENT

Patrice Laget
European Commission

Dr. Laget observed that SMEs are important to economic development in Europe and that the European Commission (EC) has tried to focus on what can be done on a community level to increase access to new technologies among SMEs. SMEs are in an increasingly competitive international economic environment but often lack the research and development (R&D) capacity to stay current with the latest technologies. In previewing his remarks Dr. Laget said his focus would be on the EC's program to aid SMEs, the result of a survey of SMEs that participate in EC programs, and how SMEs may be able to benefit from the U.S.-European Union science and technology (S&T) agreement.

The EC's program to stimulate technology access for SMEs involves three efforts to support SME R&D activity and access:

- collaborative research, in which SMEs participate and conduct R&D in a consortium setting using government, university, or private industrial labs (this a share-cost approach by which the EC provides some funding);
- cooperative research, in which two or more SMEs use EC funds to outsource R&D to a third party; and
- exploratory research, in which SMEs are given small grants for early-stage R&D.

Program Design

The EC recognizes that there are different types of SMEs with different needs. Some SMEs are very small start-ups that are developing technology but are far from commercialization. Others are technology followers, which may not innovate but need quick access to cutting-edge technologies. Finally, there are technology users, which integrate new technologies into production processes for their goods or services. Dr. Laget also observed that SMEs usually are oriented to the local level, as opposed to operating across borders. The EC is sensitive to local concerns and tries to focus on how it can have a positive impact on the business environment for SMEs.

It is important to have intelligent coordination among localities, the EC, and member states. To meet this goal, the EC, through quarterly meetings, attempts to ensure equal access among all members of the EC to information about programs to aid SMEs. In addition to providing programmatic information, the meetings facilitate communication among SMEs in the various member states.

Rules for SME Programs

To take advantage of EC technology programs for SMEs, at least two SMEs from two member states must join together to obtain grants for exploratory research (the grants are on the order of $50,000). For collaborative research two or more SMEs should be prime contractors to conduct R&D. For cooperative R&D at least four SMEs from at least two member states must identify a technology need that is then outsourced to industrial, government, or university labs. SMEs share the cost of research on a 50-50 basis, with the EC matching SME contributions up to $1 million. Large companies can be involved in these programs.

Survey Results of SMEs Participating in the Fourth Framework Programme

Of SMEs involved in assistance programs for SMEs, 70 percent were first-time participants. In cost-share (cooperative) R&D projects, 50 percent of companies are SMEs, whereas the remainder are large business firms. For SMEs receiving exploratory grants, 70 percent reported subsequent success in receiving collaborative research grants based on exploratory work.

There has been an increase in the number of SMEs participating in EU research programs. From the Third Framework Programme to the Fourth Framework Programme, the number of SMEs in collaborative research doubled. There was also a dramatic increase in the number involved with cooperative research. Cooperative research programs are used mainly by newcomers to the EU SME programs; such firms need cutting-edge technologies but do not have the resources to develop technologies themselves. Participants in cooperative R&D programs generally increase their contacts with other SMEs.

Dr. Laget identified other lessons from EU work with SMEs. It is rare, for example, for SMEs that have participated in EU cooperative R&D programs to involve themselves with local or regional cooperative R&D programs. Such SMEs seem to prefer working with the EU, and this may indicate an interest in working in the framework of the new U.S.-EU S&T agreement. Furthermore, promoting flows of information with and among SMEs is critical. SMEs report that the process of obtaining funds from the EU functions well but that there are "internal" barriers to the use of R&D generated in EU programs. In conclusion, Dr. Laget said that with the large number of SME programs in the United States and Europe it would be interesting to compare programs in the remainder of the session.

DISCUSSANTS

Jon Baron
U.S. Department of Defense
SBIR Program

Mr. Baron said that his remarks would focus on the presentation of Dr. Lerner, expand on and clarify some of Dr. Lerner's points, and comment generally on efforts to fund small business development in the United States. Mr. Baron noted that most studies of the SBIR program, whether by academics or the U.S. General Accounting Office, have taken a favorable view of the program. Dr. Lerner's work on the SBIR is regarded as one of the best such studies because it takes a systematic and empirical look at SBIR-funded firms and a comparable set of firms that received no SBIR support. Mr. Baron also expressed the belief that Dr. Lerner approached his study without any bias or preconceptions about the program. Mr. Baron then made several points to supplement the presentation of Dr. Lerner.

SBIR Versus Venture Capital

Mr. Baron pointed out that SBIR funding differs from traditional venture capital financing because it funds technology feasibility studies; this is a much earlier stage of development activity than a venture capitalist typically funds. Mr. Baron recounted the three phases of the SBIR program funding:

- *Phase I*: a six-month feasibility study that explores the likelihood that a technology may pay off;
- *Phase II*: an award of up to $750,000 that funds the development of a prototype if Phase I yields promising results; and
- *Phase III*: a matching award to Phase II recipients that is conditioned on awardees demonstrating that they have raised funds from the private sector or elsewhere in the government to develop the technology further.

Purpose of the SBIR

While noting that the broad purpose of the SBIR program is to develop technologies to help meet agency missions, Mr. Baron observed that there are differences across agencies with respect to purpose. Some agencies may place emphasis on economic development objectives in disbursing SBIR grants. For the DOD, SBIR is seen as a means to improve defense capabilities.

As an example, Mr. Baron cited the development of "savvy tag" technology as a way to better track military materials. A Silicon Valley start-up received an SBIR award several years ago to develop a radio transceiver, about the size of a deck of cards, to track military cargo anywhere in the world. Keeping track of military cargo has traditionally been very difficult, and such difficulties often result in costly waste. During the Persian Gulf War, the U.S. General Accounting Office estimated that such waste cost $2.7 billion. DOD has estimated that savvy tags could have saved $2 billion of that waste. Today, all U.S. shipments to Bosnia use savvy tags, and the tracking of cargo has improved dramatically.

Selection Challenges

An ongoing challenge to the SBIR program is weighing technological promise against business capabilities. DOD is very good at assessing the scientific merits and technological potential of an idea but is less skilled at scrutinizing business plans and assessing market potential. From internal reviews DOD has found that many companies have excellent R&D capabilities, but their ideas often do not make it to the market because of a lack of business sophistication.

Fast Track Program

Mr. Baron said that the Fast Track program had been developed to address selection challenges that DOD has faced with SBIR. Under Fast-Track procedures, applicants have a greater chance of winning Phase II funding if they have received third-party funding (e.g., from a venture capitalist or other private financing). Such financing sends a strong signal to DOD that, beyond the technical merits, the SBIR-funded technology has marketplace potential. The Fast Track program has brought a new set of companies into the SBIR program, and Mr. Baron noted the National Research Council is working on a study assessing the program.

Attilio Stajano
DGIII, European Commission

Mr. Stajano opened his remarks by complementing Dr. Laget's presentation and reiterating the importance of integrating SMEs into the EU's R&D Framework Programme. He noted that in the Fourth Framework Programme, about one-third of all research funds went to SMEs. With respect to communications and information technologies, Mr. Stajano said that the European program to promote information technology, ESPRIT, had fostered over 100,000 person-hours of cross-border R&D since 1983. This has created an environment in which scientists and industrialists, from small and large companies, have been able to develop business solutions using information technologies.

Mr. Stajano listed four conditions that are necessary for successful economic relationships between scientists, engineers, and small businesses:

- *Networks.* Creating linkages between scientists and industrialists, as ESPRIT has done, is important. Such networks must be dynamic but long lived enough to produce results.
- *Development of business solutions.* It is one thing for cooperative programs to generate scientific breakthroughs, but it is important for such breakthroughs to be translated into products or processes that help build market share for participants.
- *Ties with universities.* In citing the "Cambridge University" effect, Mr. Stajano noted the importance of developing an entrepreneurial spirit among university professors.
- *Financial support.* Early-stage financing is crucial to SME development, and Mr. Stajano said that ESPRIT has worked to provide early-stage financing to SMEs.

In concluding, Mr. Stajano said that promoting visibility of small business is another way in which governments can help. ESPRIT, working cooperatively with the U.S. Department of Commerce, plans to bring approximately 200 European SMEs to Texas in April 1999 to meet with 200 U.S. counterparts on joint electronic commerce ventures. By working to build such transatlantic connections, governments can help build vibrant small-business sectors in regional economies.

R&D in the Framework of the New Transatlantic Agenda

Kenneth Flamm, Moderator
Brookings Institution

In setting the stage for the session Dr. Flamm said that the semiconductor industry's conversion from the 200-mm wafer to 300 mm raised not only interesting technical issues but also important organizational and cultural issues for the industry.

Some might argue that the I300I initiative represents the first international effort to develop cooperatively a new equipment set for semiconductor production. Dr. Flamm said that usually a single company in the industry "bites the bullet," invests in new equipment, and debugs it at its own facility. The very large scale integration (VLSI) project in Japan in the 1970s was, however, an early example of cooperation in equipment development and deployment. The VLSI project did not deal with the entire suite of semiconductor production equipment, as I300I seeks to, but did address a large portion of it. It would be hard to dispute the notion, said Dr. Flamm, that the VLSI project contributed greatly to Japan's market dominance in semiconductors in the 1980s. One issue to consider in thinking about I300I is how unique cooperation in semiconductor equipment development really is.

Nonetheless, Dr. Flamm said that the I300I effort is the first truly international and formally organized enterprise on joint standards and tool development. The international element makes I300I distinctive and, Dr. Flamm hoped, a subject worthy of vigorous discussion.

Among the issues that I300I raises are:

- *Economic fundamentals.* As Gordon Moore pointed out in a prior session, it cost $200 million to $300 million in the 1980s to build a fabrication facility. Currently that figure is in the $1.5 billion neighborhood, making it "extraordinarily expensive to do anything in the industry." Rising costs and risks make alliances and cooperation attractive options and looking to international partners becomes a strategy to consider seriously.

- *Cooperation versus competition.* This issue is paramount to the I300I project. On the one hand, the semiconductor industry has always been very competitive, and this has driven innovation. On the other hand, cooperation looks increasingly attractive as fabrication costs continue to rise. Dr. Flamm noted an interesting pattern with respect to cooperation. Cooperation in the semiconductor industry was pioneered by the Japanese with the VLSI project in the 1970s. The ensuing Japanese success resulted in a U.S. cooperative response in the 1980s, namely SEMATECH. SEMATECH was largely a process of experimenting with the right organizational approach to make cooperation work in a U.S. setting. Just as SEMATECH was winding down as a government-industry collaboration in the 1990s, Japan, inspired by its apparent success, embarked on a new round of cooperative programs, such as Semiconductor Leading Edge Technology (SELETE), to develop tools to build 300-mm wafers, and the Association of Super-Advanced Electronics Technologies (ASET).
- *Cooperation and competition across international boundaries.* Dr. Flamm recalled negotiations between the United States and Japan on the Semiconductor Trade Agreement in the early 1990s during which Japanese negotiators argued that national boundaries were no longer relevant. The concept of borders was an "atavistic anachronism that we [the United States] had to free ourselves from." The irony today is that Japan's new cooperative efforts, such as SELETE and ASET, are basically exclusive programs for Japanese firms. In contrast, with the I300I enterprise, SEMATECH has pioneered the idea of "international technology policy cooperation," and SEMATECH has even considered allowing foreign firms into the consortium.

The crux of the national policy problem, Dr. Flamm said, is a tradeoff between technology diffusion and national advantage. Cooperation in certain areas can push the technology frontier forward at a faster pace for everyone. If that is done, however, the notion of national advantage in the control of a technology is set aside. This tradeoff could become more evident in I300I, and it has become an issue in the United States in the controversy over the extreme ultraviolet lithography consortium, an effort by several U.S. companies to use technology generated by U.S. national labs in conjunction with Japanese companies to advance lithography technology. In concluding, Dr. Flamm recalled from his experience at the U.S. Department of Defense (DOD) the attitude toward foreign suppliers. He said that DOD operated under three unwritten rules in considering work with foreign suppliers:

- *Monopoly.* DOD did not want a key component monopolized by a firm

from one country. If there was a diverse source of suppliers around the globe, DOD saw little problem with using foreign suppliers.
- *Plausible threat of supply cutoff.* If a choice was made to use a single foreign supplier or a set of foreign suppliers, is there a plausible threat of being deprived of a key component? If such a threat existed, DOD would prefer not to use foreign suppliers.
- *Technical advantage.* In some cases, DOD wants to be better than anyone else in a certain technical area but realizes that it cannot be the best in all areas. When the DOD wants strategic superiority in a technical area, it might prefer a U.S. supplier. If that is not the case, it might welcome cooperation with international partners.

In turning the program over to William Spencer of SEMATECH, Dr. Flamm observed that business may have a similar set of "unwritten rules" when considering whether to engage in international cooperation.

THE 300-MM INTERNATIONAL INIATIVE

William Spencer
SEMATECH

Dr. Spencer said that his remarks would focus mainly on the process of bringing the I300I program together, as opposed to the technical details of the conversion from a 200-mm wafer standard to a 300-mm standard. He agreed with Dr. Flamm that the I300I initiative represents a unique form of international collaboration, and Dr. Spencer emphasized that the collaboration was on technology, not basic science. I300I is a bottom-up initiative of private companies and receives no funding from any government. Dr. Spencer also noted that I300I focuses on precompetitive technology for member firms, although what is precompetitive for Gordon Moore at Intel is very competitive for a semiconductor manufacturing equipment supplier such as John Shamaly at the Silicon Valley Group.

Lessons from SEMATECH

In developing I300I, which is a separate subsidiary of SEMATECH, management decided to stay away from intellectual property issues. This was a lesson from SEMATECH, which has filed relatively few patents in its 10 years of existence and has no trade secrets. The movement of people has been an important contributor to SEMATECH's success; SEMATECH has over 200 member-company employees, known as assignees, in its work force. For I300I about 50 employees from member firms are assigned to the project. Using people as the main tool for technology transfer will be as prominent in I300I as it has been in SEMATECH.

Importance of the Semiconductor Industry

Dr. Spencer then turned to the topic of semiconductors in everyday life and observed that he would cover some of the same ground that Gordon Moore had in a prior session. The typical automobile today contains more computing power than the lunar module that landed on the moon nearly 30 years ago. Early communications satellites, a technology Dr. Spencer worked on, had approximately 2,000 transistors and orbited about 100 miles above earth. Today, communications satellites orbit at 200 times the distance and contain roughly 100 times the number of integrated circuits as the first satellites had transistors. The technology that enables these advances is the semiconductor. The semiconductor industry supports a worldwide electronics industry whose sales are approximately $1 trillion, with 15 to 20 percent of that being semiconductor sales. Over time, sales in the semiconductor industry have grown rapidly, as the performance of devices has improved. In addition to the growth of semiconductor industry sales, the equipment and materials firms that support the semiconductor industry have grown quickly.

Dr. Spencer stressed two points about the future growth of the industry. First, if the semiconductor industry is to adhere to the pattern of making things "faster, smaller, and cheaper," technology in the semiconductor equipment sector must advance. About 75 percent of future improvement in device functionality would come from better manufacturing equipment, with the remaining improvement coming from changes in the structure of the transistor and other design advances. Second, future growth would depend on transition to a 300-mm wafer, but the slump in industry sales beginning in 1996 and continuing through 1998 has affected the I300I initiative. There are fewer resources available for wafer conversion. Even with the current slump, long-term growth in the industry is expected to be 15 percent, approximately the historic growth rate.

Conversion to 300-mm Wafers

In discussing the change from 200- to 300-mm wafers, Dr. Spencer stressed the improvements in functionality of the semiconductor. In terms of cost per function, the semiconductor industry has been advancing at a rate of 25 to 30 percent per year for the past 30 years. That is, the combination of more circuits per chip and less expensive chips has yielded productivity improvements approaching 30 percent per year for 30 years. Important drivers to this productivity have been larger wafers and better interconnection technology between circuits.

When the industry retooled for 200-mm wafers, moving from 150 mm, one U.S. firm purchased the full suite of tools for 200-mm fabrication and paid for much of the tool development. When it was suggested to that company that it do the same thing for conversion to 300 mm, the response was emphatically negative because of the huge cost of doing so. In fact, continued Dr. Spencer, it has been estimated that the industry-wide cost of converting to 300-mm wafers is between $10 billion and $20 billion. This does not include the $2 billion cost for building a single fabrication facility. Dr. Spencer also remarked that the United States is currently dependent on either European or Japanese suppliers for the silicon wafers from which semiconductor devices are fabricated.

National Technology Roadmap for Semiconductors

An important part of the I300I cooperative process is the industry's main planning tool—the National Technology Roadmap for Semiconductors (NTRS). The NTRS was first published in 1994 and has been updated every two years since. Dr. Spencer observed two recent important developments with respect to the roadmap: (1) it will become international in 1999 as engineers from all over the world will be asked to participate and (2) it will be updated on an ongoing basis; as technology changes, the roadmap will be revised, with revisions available on-line (already, the NTRS is available on the World Wide Web at http://www.sematech.org). The roadmap does not propose technical solutions but specifies capabilities that the industry must meet and the time by which they must be met, for historical productivity rates to continue.

Issues in the Conversion to I300I

Conditions in the world economy and the semiconductor industry have caused a delay in attaining 300-mm goals. The initial goal, which was widely publicized, called for a 300-mm pilot line to be completed by 1998, using 180-nm line widths on chips. However, the semiconductor industry has had a year of unexpectedly poor sales, largely because of the Asian economic crisis. Combined with the overcapacity among chip makers, there has been a glut of dynamic random access memories (DRAMs) and a drop in prices that has squeezed profits. Dr. Spencer noted that as of early June 1998 the price on the spot market for 16-megabyte DRAMS was $1, whereas the price for 64-megabyte DRAMS was $6; both levels are below production costs. Such tough economic conditions in the industry have caused a delay in investments to convert to 300 mm.

The problems have created repercussions for semiconductor equipment manufacturers. The equipment makers were trying to meet the 1998 deadline and indeed were anxious to do so to have their equipment adopted widely in fabrication facilities. Now, instead of a 1998 deadline for 300-mm wafers using 180-nm line widths, equipment manufacturers are being told that the deadline is 2001, with 150-nm technology instead. This makes investment planning very difficult for equipment makers, and further delays could negatively impact the finances of semiconductor equipment firms.

The I300I Project

Members of I300I are Hyundai, Philips, STMicroelectronics, Siemens AG, and the Taiwan Semiconductor Manufacturing Corporation. Dr. Spencer said that the Asian financial crisis has prompted two Korean firms not to participate in I300I. The reason the U.S. semiconductor industry has sought international cooperation, Dr. Spencer continued, is because the business is global. Any semiconductor company must be able to build a factory anywhere in the world and draw on equipment vendors worldwide. In fact, much of the equipment for 300-mm production will come from European and Japanese firms. If an equipment firm involved in I300I were to be prohibited from selling globally, it could not be a true leader in the industry, and it would not have the resources to invest in research and development.

When I300I was being formed, every semiconductor company in the world was invited to participate, from Europe, to Japan, to Korea, Singapore, and Taiwan. Japanese firms declined to participate, so there are two parallel I300I efforts in the world. The Japanese effort, called SELETE, includes 10 of Japan's largest semiconductor firms.

Keys to Success

Dr. Spencer listed a number of organizational and technical factors that will be crucial to I300I's success.

Assignees

Member companies send assignees either to I300I's headquarters in Austin, Texas, or to facilities of equipment manufacturers for tool development. Some of the best engineering talent resides at equipment manufacturers, and it makes sense to develop and debug equipment with the engineers of equipment manufacturers. This is distinct from the Japanese approach to 300-mm conversion, which has a more centralized structure.

Standards

At one time it looked as if there would be two sets of standards for 300-mm development. It was a major accomplishment of the I300I project, working with the Semiconductor Equipment Manufacturers Institute, to reach agreement with Japan on a single set of standards. Among the many standards to be settled are flatness, alignment, and impurities.

Automated Materials Handling

Unprocessed wafers are expensive, about $1,000 to $1,200 each, and with integrated circuits on them they are even more valuable. Twenty-five 300-mm wafers with Intel processors etched on them are worth $1 million. It is therefore important to automate the handling of wafers to prevent costly breakage.

Cost

The I300I project must develop a 300-mm process technology that is initially no more than 20 percent more expensive than 200-mm technology. Otherwise, the conversion to 300 mm is not worth it for the industry. This cost target must be met across more than 200 pieces of equipment.

Return on Investment

Participating companies must receive a return on investment of about two to three times what they invest in I300I. Meeting that goal will require getting the best people from member companies as assignees. Dr. Spencer recalled that SEMATECH had problems in its early days in recruiting people from member companies to serve as assignees. That situation is now reversed, as SEMATECH usually has many more applicants than slots for assignees and, moreover, reserves the right to send substandard assignees back to the member companies. Such practice will be carried forward in the I300I project.

In conclusion, Dr. Spencer said that SEMATECH and I300I have shown that fierce competitors can come together to cooperate on precompetitive technologies. It requires a commitment of resources, people, and time, but such cooperation also promises large payoffs.

DISCUSSANTS

John Shamaly
Silicon Valley Group, Inc.

Mr. Shamaly began his comments by observing that the I300I project is about productivity for the semiconductor industry. A simple calculation shows that, with the same throughput of wafers, switching from 200- to 300-mm wafers would increases productivity by 2.25 times. In conjunction with shrinking line widths, in accordance with Moore's law, productivity would grow even more. Finally, the Silicon Valley Group (SVG) sees improvements in step-and-scan technology as fueling additional productivity advances. Together, Mr. Shamaly said that these technological changes could increase industry productivity by 10 times in the next five to seven years. He observed that the demand for chips would have to keep pace with these changes to support the sales of semiconductor manufacturing equipment, which in turn fund tool development.

With respect to cooperation, Mr. Shamaly said that, as a manufacturer of automated wafer processing equipment, SVG serves a worldwide market and thus favors global cooperation when appropriate. There are guidelines that SVG follows in considering cooperation with other firms:

- adherence to U.S. government regulations with respect to cooperation with foreign firms;
- keeping its obligation to stockholders and employees, namely, to earn enough profits to maintain or increase employment at SVG, while stockholders receive a good return on their investment; and
- maintaining competitiveness for SVG and the entire equipment industry. Increased cooperation, along with productivity increases in the semiconductor equipment sector, may limit the size of the world market. Much depends on the ability of worldwide demand for semiconductors to keep pace with productivity improvements.

In summary, Mr. Shamaly said that SVG favors cooperation in principle and recognizes its potential benefits. However, SVG is concerned about overcapacity in the industry and the negative economic consequences it has in the form of low chip prices. As with other productivity-enhancing efforts in the industry, Mr. Shamaly concluded, international cooperation must take place in the context of growing chip demand.

Robert Hance
Motorola

Mr. Hance said that his remarks would focus on another collaborative effort in the industry, specifically one that is designed to address materials issues in semiconductor fabrication. Before discussing this, Mr. Hance noted that Motorola is a member of I300I, continues to participate in SEMATECH, and has benefited from its participation in both consortia.

As the semiconductor industry switches to 300-mm wafers, Mr. Hance said, there is also the challenge of improving the materials that go into semiconductor fabrication. To address this challenge, the European Commuission has founded a consortium called the Hector 300-mm Project, which stands for High Epsilon Materials Cluster Tool for Optimized Rapid Deposition of stacked capacitors on 300-mm wafers. The Hector 300-mm Project is managed by Directorate General III (DG III) and related to the European Union's information technology research and development program, ESPRIT. In effect, this effort adds value to the 300-mm conversion project by making sure that appropriate materials are available by the time 300-mm manufacturing processes are on-line. Any time a change in wafer size is made in the industry, Mr. Hance said, new materials must be developed to mirror new manufacturing processes.

Motorola is able to participate in this European consortium because of its manufacturing plants in Europe. The consortium began operation in March 1998, is expected to last 28 months, and will receive 10 million ecu, or $11 million (U.S.), from DG III. The members of the consortium are Aixtron, ASM Lithography, Steag-AST, Siemens, Motorola, Lucent, Austria Mikro Systeme, the Fraunhofer Institutes, and RC Juelich.

In concluding, Mr. Hance stated that the European effort was in tackling 300-mm issues at the "module" level rather than the global level at which I300I operates. Both efforts are necessary for the industry to take full advantage of the switch to semiconductor fabrication on 300-mm wafers.

Michael Borrus
University of California at Berkeley

Because the prior discussants, in addition to William Spencer, put so much content on the table in terms of international collaboration, Mr. Borrus said that he would concentrate on the context of international collaboration. Cross-border collaboration is not only useful for promoting innovation but increasingly essential as the economy becomes more global. International collaboration is part of a nascent effort to globalize the innovation process. Mr. Borrus said that he would discuss three reasons for the growing globalization of the innovation process.

The Growing Cost and Risk of Innovation

Growing costs and risks were mentioned by Dr. Flamm and Dr. Spencer and are part of what Mr. Borrus termed the *Business Week* image of innovation. That is, the need for quick time to market within firms, narrow market windows, and costly research and development (R&D) creates incentives to look worldwide for partners to meet these challenges. In this view we have a borderless world in which multinational firms work together to push the frontiers of innovation.

Increasing Specialization of Technical Skills

A consequence of the globalization of innovation has been that regions maintain and strengthen their hold on unique technical skills. Mr. Borrus recalled an anecdote regarding the sale by the American conglomerate ITT of its German subsidiary Standard Electric Lorenz (SEL) to the French company Alcatel. Mr. Borrus asked a German friend if there was any worry over the sale; his friend responded that "SEL was German when it was American and will be German when it is French." The story underscores how a company or a region's innovative character is "sticky"; it is embedded in the people, the network of suppliers, and local institutions (such as universities) in a particular region.

Mr. Borrus continued by identifying areas in which such "stickiness" is apparent. Precision engineering skills reside largely in Europe and Singapore; certain materials technology thrives in Japan and Europe; software development and chip design are very strong in the United States. Such global exploitation of know-how results in certain places being the home for certain types of know-how. For this perspective on innovation to take hold, cross-border collaboration is essential.

Mr. Borrus emphasized that in this image of innovation international collaboration does not result in greater diffusion of technology but greater specialization of skills in specific regions. This suggests that further globalization will increase specialization and product differentiation, not result in more homogenization. If this is true, the tradeoff between global benefits of collaboration versus national economic advantage, raised earlier by Dr. Flamm, is a false one. We can, argued Mr. Borrus, increase the rate of innovation while reinforcing national advantages in areas of specialization.

Standard Setting and Local Reinvestment

Mr. Borrus argued that the launch market for an innovative new product is typically local, and that if such a product sets an industry-wide standard, the local economy benefits. When a product becomes a standard, additional investment in a specific region results as production volumes grow and as a network of supporting industries and suppliers grows. The products of companies such as Microsoft and Intel in the United States and SAP, Nokia, and Ericsson in Europe are examples.

With half of the world's economy in North America and Europe and with an even greater share of high-technology markets, U.S.-European cooperation is a way to ensure that standards for new high-technology goods are set by Europe and the United States. The economic benefits of standards setting will, said Mr. Borrus, then accrue to the United States and Europe.

In summing up, Mr. Borrus reiterated the essential role that U.S.-European collaboration must play to commercialize products and reap the fruits of R&D and innovation. There is, however, political resistance to this idea in some quarters in the United States. He noted the opposition among some members of Congress to the participation of foreign companies in the extreme ultraviolet lithography consortium. If there are balanced contributions from both sides of the Atlantic in cooperative projects, Mr. Borrus said that such political problems can be avoided. In the end, U.S.-European cooperation constitutes a huge opportunity for both regions to capture substantial economic benefits in high-technology markets.

COMMENTS FROM PANELISTS AND THE AUDIENCE

William Spencer commented on John Shamaly's account of the Hector 300-mm Project by observing that there is a substantial amount of government support for small semiconductor equipment companies outside the United States. He noted that such governmental financial support for small equipment companies does not exist in the United States, and this could eventually hurt the U.S. market share in this sector.

Kenneth Flamm responded to Mr. Borrus's points about a tradeoff between global technological benefits and national economic advantage by agreeing that there were cases in which there was no such tradeoff. Dr. Flamm argued, however, that it is an empirical question as to whether the tradeoff generally does or does not exist. Efforts to convert to 300-mm wafers are an example. Japan has decided to have an exclusive 300-mm conversion project, apparently hoping to reap the economic benefits in the manufacturing equipment sector. Semiconductor device makers in the United States and Europe have agreed in I300I to move equipment development forward based on common interests and to compete in the market for devices.

Dr. Flamm also raised a question about intellectual property. Even though Europe has accelerated its economic and political unification and even though the European Union (EU) has a common patent agency, Europe still does not have a single forum for litigation of patent disputes. Dr. Flamm solicited input from the audience as to whether his understanding was correct about the status of dealing with patent disputes in Europe.

In response, Dr. Jorma Routti, Director General of DG XII of the European Commission, acknowledged a need for harmonization between the United States and Europe on patent law in the context of the World Intellectual Property Organization. This, however, was outside the scope of the U.S.-EU science and technology agreement. Dr. Routti expressed the hope that patent law would encourage the publication and dissemination of scientific research and preserve the ethic of openness in scientific inquiry.

Dr. Routti also inquired about the balance of software and hardware development in the development of semiconductor devices such as integrated circuits. Specifically, he wondered whether there was a software or hardware tradeoff or a proper balance between the two that researchers or policymakers should explore.

Dr. Spencer responded by describing the resource tradeoffs that semiconductor makers face in developing products. They can put money into the development of competitive products or into manufacturing processes and tool development. Companies must, of course, fund some of both activities, but the purpose of projects such as SEMATECH and I300I is to allow companies to minimize funds for the latter. This is why process and tool development efforts are known as "precompetitive" projects. The result is relatively more funds available for chip design, which is the main area in which semiconductor companies compete. Such design activity encompasses development of software and hardware applications for the chip, and individual companies will place different emphases on each element, depending on their design approaches. This, in fact, is very much a competitive and proprietary issue for semiconductor companies.

Alan Tonelson of the U.S. Business and Industrial Council Educational Foundation asked Mr. Borrus about the tradeoff between global technology dissemination and national economic advantage. Mr. Tonelson wondered in particular whether this work contradicted earlier work of Mr. Borrus and his colleagues at the Berkeley Roundtable, which argued that there has been a convergence of manufacturing capabilities in Asia. Mr. Tonelson suggested that there has been an "electronics traffic jam" in Asia whereby too many countries have been making too many of the same electronic goods such as DRAM chips. Such overcapacity in electronics manufacturing, said Mr. Tonelson, may be at the root of the Asian financial crisis.

Mr. Borrus responded by noting that, whereas collapsing DRAM prices may have contributed to financial problems in Korea, it would be very difficult to assign a causal link between falling DRAM prices and the general economic crisis in Asia. Moreover, Asia does have a well-defined division of labor that supports the notion that a homogenization of technical tasks is not occurring. Mr. Borrus cited software development in Bangalore, process know-how in Singapore, and digital design in Taiwan as examples of Asia's division of labor.

Internationalization of the Technical Workforce and Transatlantic Cooperation in R&D

William Wulf, President
National Academy of Engineering

In setting the stage for the panel's discussion on internationalization of the technical work force, Dr. Wulf commented on the importance of people in making transatlantic cooperation work. The ability of people from different cultures to work together will be the key to transatlantic collaboration, which makes internationalization of the technical work force an issue to be explored. People who are able to work well with individuals from different cultures will prosper in the global economy. Dr. Wulf expressed the hope that the panelists would discuss how to improve links between the technical work forces of the United States and Europe.

E. Praestgaard
European Science and Technology Assembly, Denmark

Dr. Praestgaard began his remarks by stating that it should be possible to develop a "win-win strategy" to further the internationalization of the technical work force while respecting the cultures and autonomy of individuals and nations. It is a matter of political will to develop and implement such a strategy. After all, Dr. Praestgaard said, "we are a generation of European scientists trained in the United States," and this should have a positive impact in creating a truly international technical work force.

Although the flow of Europeans to the United States for scientific and technical training remains strong, there is a shortfall of U.S. scientists and engineers in Europe. Dr. Praestgaard recalled a time when many U.S.-trained scientists did postdoctorate work in Europe and remained there. This practice has been on the decline in recent years. Dr. Praestgaard cited two reasons for this downward trend:

- *Funding.* Currently a good deal of funding goes through government programs on both sides of the Atlantic and is therefore directed to specific research areas. This has become a substitute for "professor-to-professor" contacts that once served as a catalyst for many informal postdoctorate programs, especially for U.S. students coming to Europe.
- *Career planning.* There is a perception that choosing to pursue a postdoctorate in Europe is a bad career move for young American researchers. Even if that perception is ill founded, it has had a noticeable impact on the flow of American researchers to Europe.

In looking at the European experience, Dr. Praestgaard said that the European Union Framework Programmes have been successful in internationalizing research by explicitly constructing teams of researchers from different countries to work on different projects. For U.S.-European scientific collaboration, the key to long-term success will rest on a steady exchange of scientific and technical personnel. It is important to develop mechanisms for such exchanges.

Mechanisms for Exchange

Dr. Praestgaard said that many European Union members may view the exchange of scientists and engineers as a worthy goal but one that is best left to bilateral efforts. Bilateral measures may work for large countries, such as the United Kingdom, Germany, and France, but they are likely to be less helpful for smaller countries such as Denmark. European-wide efforts would therefore be useful to ensure that all countries benefit from exchanges of scientists and engineers. Such European-wide efforts could also battle the perception that collaboration with Europe is held in low esteem in the United States; by marshaling all of Europe's resources, collaboration may be more attractive for Americans.

Developing Networks

One way to develop rich networks between the United States and Europe is to have researchers from both sides work together on projects that are two to four years in duration. Dr. Praestgaard said that the European Commission already has mechanisms for developing intra-European networks. Extending such mechanisms to the United States and other countries could be one way to develop long-term networks, although Dr. Praestgaard cautioned that there may be concerns about the appropriateness of opening up such existing European mechanisms to other countries.

Need for Funding

As important as it is to ensure that individuals from both sides engage in greater interaction, Dr. Praestgaard said that more funding is crucial to further the internationalization of the scientific and technical work forces in the United States and Europe. An important objective is increasing the flow of postdoctorate researchers coming from the United States to Europe. For that to occur, Dr. Praestgaard concluded, more money is needed.

H. Glatz
DaimlerBenz, Germany

Mr. Glatz said that he would focus on education and its role in promoting transatlantic cooperation as well as how the global market will promote internationalization of the work force. First, he wanted to make some points about intellectual property rights, especially in light of Dr. Routti's comments in the prior session. Mr. Glatz said that the different approaches to patents in the United States and Europe—the United States having a "first to file" approach versus Europe's "first to invent" approach—is the key difference between the two sides regarding R&D. Mr. Glatz stressed that the European Patent Agency is a simplifying mechanism, because it does not require patent applications in 15 countries. Even though different intellectual property rights laws remain a barrier in conducting transatlantic R&D, the business community and the governments on both sides, as well as academics, are working to lower the barriers that are associated with intellectual property law.

Education

Promoting the internationalization of the scientific and technical work force through educational institutions remains a key theme in Europe, but the transatlantic component has been neglected. Mr. Glatz sees many U.S. students working in his company, but he believes that the number of such students has been declining in recent years. Less money for such exchanges is the main reason for the decline.

The Rise of Global Markets

For business the rise of global markets closely parallels the growing internationalization of the scientific and technical work forces. If Daimler Benz is developing a car in Alabama, many of its German engineers, along with employees from its Palo Alto research and design center, will spend time on site. As the Daimler Benz–Chrysler merger proceeds, the internationalization of the work forces from both companies will accelerate.

In light of the trend toward internationalization, Mr. Glatz asked whether the Japanese "closed-door" strategy in developing the 300-mm wafer technology was sound. Alternatively, was it better to use the "open-door" approach as

SEMATECH has done in I300I that invites wide participation? Mr. Glatz noted that Japan has had some success in the past with the "closed-door" strategy. Whatever the merits of such a strategy, it surely will not promote the internationalization of the scientific and technical work forces.

Mr. Glatz concluded by saying that joint work among young researchers on international projects acclimates researchers to working in an international context. This is the key to building future links for collaboration. Not only do such links help business, but they further foster the internationalization of the technical work force.

Henri Conze
Ministry for Defense (1993–1996), France

Mr. Conze placed his remarks on the internationalization of the technical work force in the context of the growing importance of transatlantic cooperation to the European economy.

Industrial Restructuring

With European economies becoming increasingly integrated, European industry will need to continue restructuring to meet competitive challenges. Growing transatlantic trade will also affect European industrial restructuring. Greater transatlantic S&T cooperation will be a part of growing trade relations, which means that S&T cooperation will likely play a role, perhaps indirectly, in Europe's changing industrial structure.

Role of Individuals in Making Cooperation Work

Mr. Conze recalled that during his career in the French government he signed 20 cooperative agreements between the United States and France. Only one agreement is still alive; the rest failed, mostly because of bureaucratic inertia. Only committed individuals can make such agreements work, and this will be true for the U.S.-EU S&T agreement, too. International agreements, when they fail, usually fall victim to cultural misunderstandings. Individuals must work through such misunderstandings, and one way to build the capacity to do this is through long-term relationships among S&T professionals. The U.S.-EU S&T agreement is an opportunity to pave the way for a world in which American and European researchers can operate together in the global marketplace.

In concluding, Mr. Conze said that resistance to collaboration among scientists and engineers still exists on both sides of the Atlantic. Engaging students in collaboration at an early stage in their careers is one way to overcome such resistance. For both sides, such integration of scientific and technical work forces will increasingly be a condition for competitive success in global markets.

Gary Poehlein
National Science Foundation

In previewing his remarks Dr. Poehlein said that, in addition to his position at the National Science Foundation (NSF), he is a professor of chemistry at Georgia Institute of Technology, whose main campus is in Atlanta, but which also has a campus in France. He stated that his comments would reflect his experiences at NSF and Georgia Tech.

During his academic career, Dr. Poehlein has taught many foreign students, many of whom have stayed in the United States and contributed greatly to this country. He has also seen a number of U.S. students spend time abroad but said that the United States would benefit from having more students spend a greater amount of time overseas. Dr. Poehlein identified the following barriers to U.S. science and engineering students studying abroad:

- *Finances.* More fellowships are needed for overseas study.
- *Language.* Language barriers remain significant, and language training would help lower the barriers.
- *Impatience of young people.* Many students do not want to extend their educational programs by one year through overseas experience.
- *Lack of perceived career advantage.* The bias against spending time abroad exists, but it may be declining, especially as the chemical industry becomes more international.

Dr. Poehlein observed that there was a large difference in perspectives on international study between business schools and the science and engineering community. Business students see a clear advantage to international experience, but science and engineering students do not widely share this view. Because the business community is ahead of its counterparts in science and engineering on international education, Dr. Poehlein suggested enlisting the business community to advocate international education among scientists and engineers. Business wants to hire the best science and engineering students from universities, and it also wants to engage universities in research. Perhaps business could deepen its relationships with scientists and engineers at universities by providing internships for students abroad. Students could, for example, work at overseas industrial labs, thereby addressing students' financial concerns while providing business something of value.

In conclusion, Dr. Poehlein said that to advance the internationalization of the technical work force U.S. business should offer more opportunities for U.S. science and engineering students to gain experience abroad.

Dieter Seitzer
Fraunhofer Institute, Erlangen-Nürnberg, Germany

In framing his remarks, Dr. Seitzer gave some historical background on the Fraunhofer Institutes, saying that Fraunhofer himself was an optical scientist. Fraunhofer was also an engineer who developed precision tools and glass with the objective of doing better research in optics. Eventually Fraunhofer became an entrepreneur, starting his own business to manufacture glass and precision tools for optics.

Fraunhofer's background very much captures the spirit of the institutes in Germany that bear his name. Science, engineering, and entrepreneurship are all closely tied together in the Fraunhofer Institutes, which try to do cutting-edge research and engineering while also working closely with industry so that the institutes' research can be applied in manufacturing and business processes.

To be close to their customers, the Fraunhofer Institutes are spread out around Germany, usually close to industrial centers and universities. Of its 9,000 employees, about 3,000 work part time—these are mostly students. Often, students start out at Fraunhofer as assistants; as they advance to Ph.D. studies, they become valuable contributors to the institutes' work. Of Fraunhofer's full-time technical staff, totaling 4,500, nearly two-thirds also hold positions with universities.

Much of the institutes' work is contract research and development for industry; most studies look at technological feasibility or develop prototypes. The Fraunhofer Institutes offer continuing education to scientists and engineers in industry who may have graduated 15 to 20 years ago and need to be updated on the latest developments in their fields.

Finally, the Fraunhofer Institutes have exchange programs with other European countries. Dr. Seitzer said that Fraunhofer participates in the European Union's COMET program by which students in one country do research and engineering work for businesses in another country. With over 200 industrial clients, Fraunhofer has a great capacity to place students in German industry. Dr. Seitzer concluded by saying that the Fraunhofer Institutes are ready to welcome U.S. students through exchange programs.

COMMENTS FROM THE AUDIENCE

Brian Randall of the University of Newcastle suggested that a prestigious fellowship program be established to encourage more U.S. students to spend time in Europe at university or business research institutions. He said that such a fellowship program could be an adjunct to the U.S.–EU science and technology agreement. In light of the National Academy of Sciences' New Vistas conference and the Einstein statue outside the Academy's main building, perhaps such a program could be called the Einstein Fellowship, which would capture the transatlantic objective of the fellowship. It would be important, Dr. Randall concluded, to use prestige and funding as an incentive to encourage U.S. science and engineering students to work and study in Europe.

Concluding Remarks

Jorma Routti
Director-General DGXII
European Commission

The statue of Albert Einstein outside the National Academy of Sciences' main building in Washington, D.C., reminds us of the search for the simple and beautiful laws of nature. Einstein's insight on the relationship between energy and mass is a good example of such scientific simplicity and elegance. On the other hand, a quotation credited to Einstein—"Everything should be made as simple as possible but not simpler"—cautions us about oversimplification. It is useful to keep this idea in mind when we seek solutions to the complex problems of today's world.

The study of complex phenomena has progressed rapidly in recent years. Fractals, bifurcations, and chaotic phenomena are found in mathematics, weather patterns, biology, and economic theories. The work of Belgian Nobel Prize winner Ilya Prigogine has contributed greatly to the understanding that seemingly simple things, even at the atomic level, have great uncertainties. Prigogine's recent works, *The End of Certainty* and *The Laws of Chaos*, shed light on the inherent complexity of the early universe. To improve the understanding of complex phenomena, the Santa Fe Institute in New Mexico brings together many Nobel Prize winners to explore complexity. Researchers in Santa Fe have found that our world is not predetermined, that it is far more unpredictable than we have imagined and hence much more challenging than the deterministic Newtonian world. In the economic realm, complexity theory points to the difficulty in predicting the economy's evolution and that small changes in economic conditions can lead to swift and dramatic changes in market shares within industries. The dynamism of knowledge-based companies in the communications, electronics, and biotechnology sectors is a manifestation of complexity in the economy. The end result is the growth of opportunities for entrepreneurs, who may be able to reap large rewards in a fast-changing, though risky, economy.

As we all know, science and technology are among the principal driving forces of the world today. They form the foundation of knowledge-based industries, and they are needed to develop solutions for complex problems facing our societies. *Science, The Endless Frontier*, the classic book by Vannevar Bush (1945), has defined science policies in the United States for decades. In 1997 the European Commission published *Society, The Endless Frontier*, which analyzes links between science and society and defines the challenges of today and tomorrow. The United States and Europe should draw lessons from each of these important documents so that science on both sides of the Atlantic complements each other in the best possible way.

The new approach, from *Society, The Endless Frontier*, is also the basis of the new European Union's Framework Programme for Research and its problem-driven structure. This approach should interest our American colleagues because the framework's key recommendations open wider access to European research while asking for complementary American contributions. The important feature of such science collaboration is that it is not a zero-sum game where one party wins at the expense of the other. Rather it can lead to win-win results in many areas of common interest, on the basis of reciprocal contributions and mutual benefits.

We have discussed many issues and technologies during this conference. To draw conclusions from the rich program is not an easy task. Specific recommendations and conclusions have already been reported from the parallel workshops. So my conclusions are of a more general nature, and I will summarize them in eight points.

1. *High level of interest.* We can fairly say that the new U.S.-European Union (EU) science and technology (S&T) cooperation agreement has generated a lot of interest on both sides of the Atlantic. The agreement is an instrument to promote scientific cooperation across the Atlantic. This legal and administrative framework should be used in a proactive way.

2. *Broad-based involvement.* Although public authorities and their agencies will play a very active role, a top-down approach alone is not sufficient. Industry (large and small), academia, and individual laboratories and researchers must actively engage themselves in joint projects in areas of common interest. We should use our agreement as a tool for efficient cooperation to avoid unnecessary and costly duplications on each side.

3. *Use of advanced communications technologies.* Building timely and easily accessible information channels and using modern high-technology information and communication are of the utmost importance for efficient implementation of the agreement. We need to make all interested researchers and

policymakers aware of the potential that this agreement offers. Otherwise, there is a risk that it will remain a skeleton.

4. *Continue to develop priority areas for cooperation.* The reports from our conference sessions on the priority areas indicate that there are topics to be explored further. This should be done soon, and indeed efforts are already under way. On the European Union side, the Fifth Framework Programme will be quite well synchronized with the U.S.-EU S&T agreement in the beginning of 1999. However, those that are directly concerned in the already-selected priority areas should begin to launch concrete collaborative actions now. The EU will continue the process of selection of further priority areas after this conference. I also believe that the EU's new approach to research, with its focus on finding solutions to major problems facing society, is of great interest to the U.S. side.

5. *Maintain momentum for cooperation.* A follow-up to this conference will held in Europe in 1999, and at that time we will continue to explore areas for cooperation. Some areas will be jointly selected during an upcoming informal meeting of the Joint Consultative Group here in Washington and the forthcoming formal meeting in Brussels. In the meantime the momentum generated by this first conference should not be lost. Contact persons designated by each side shall pursue the follow-up jointly with their colleagues responsible for S&T collaboration in the European Commission and in the U.S. government agencies.

6. *Continue bilateral cooperation.* With increasing S&T cooperation between the European Community and the United States, we should neither forget nor underestimate the numerous opportunities for collaboration at the bilateral level. Collaboration can proceed between the United States and EU member states or directly between universities and industries from both sides. Hence, it is important to use the U.S.-EU S&T agreement in a selective and intelligent way and to respect what we call "subsidiarity" in the EU. Let us choose the most efficient level and channel of cooperation and avoid unnecessary duplication.

7. *Engage all types of businesses.* Effective action also needs industrial partners, including small- and medium-sized enterprises (SMEs). We must not forget that SMEs have an important role to play in all sectors of technological development. The role of SMEs in transatlantic cooperation opportunities should be emphasized.

8. *Engage young scientists and engineers.* We should involve our young scientists and engineers in transatlantic cooperation. This is a very cost-effective way of building international collaboration. It will make young researchers aware of the possibilities for cooperation as well as the challenges inherent in transatlantic relationships that can be both complementary and competitive.

On behalf of the European Commission, let me express our sincere thanks to the National Academy of Sciences and the National Research Council for the excellent arrangements for our meeting. Chairpersons and rapporteurs also deserve our thanks for summarizing the contributions in the workshops. And all of us have enjoyed the excellent speakers from both sides of the Atlantic.

I also want to thank the United States for its tradition of opening up its universities to students from other countries. My country, Finland, has been fortunate enough to have an aggressive program of sending students to the United States. I want to express my personal thanks for the opportunity I had in the United States; many others from Europe have similarly benefited from U.S. openness.

In closing I want to talk about the brain as a model for collaboration between the United States and the European Union. Gordon Moore's speech reminds us of the increasing miniaturization of computing power, as reflected in Moore's law. Ongoing efforts, such as I300I among semiconductor manufacturers, will make smaller and more powerful chips. Before too long, shrinking chip size could mean that electronic circuitry could approach the size of neurons in the brain.

The brain is still better than a computer because of superior pattern recognition. For example, when we notice a person we have seen before, our brain takes one-tenth of a second to tell us whether we know that person. With each neuron connected to 10,000 other neurons by synapses, our brain can make this calculation. Although today's neural networks can do some truly astounding things in completing complex tasks, not even the most advanced supercomputer today can accomplish pattern recognition with the speed and accuracy of the brain.

Why can the brain work so quickly? It works so fast because it must: millions of years ago, if we did not quickly recognize a lion on the prairie, we would have been killed. Today, if we cannot recognize the truck coming around the corner at us, we are in danger.

Lots of connections make the brain work so swiftly and effectively. The model of the brain is also the best model for scientific collaboration. Fostering many connections between European and U.S. scientists will be the key to making increased transatlantic S&T collaboration fruitful for both sides. Collaboration is increasingly the model for the economy today in which small research organizations team with large companies for production and marketing. Like the brain, however, scientific collaboration needs many neurons and a large number of connections.

To take another analogy from computer science, Control Data Corporation made computers in the 1960s with some of the most powerful processing capabilities of that time. But the computer was connected to "dumb terminals" and thus greatly limited in its scope of use. Today, powerful computers involve interactive networks of intelligent workstations. That is what we need today in our economies and our approaches to scientific work-intelligent networking.

Selected Bibliography

Alic, John, et al., 1992. *Beyond Spinoff: Military and Commercial Technologies in a Changing World.* Harvard Business School Press, Boston, Mass.

Amsden, A.H., 1989. *Asia's Next Giant: South Korea and Late Industrialization.* Oxford University Press, New York.

Archibald, R. B., and D. H. Finifter. 1998. "Perspectives on the evaluation of the SBIR program with an application to the NASA Langley Research Center." Paper presented at the American Economic Association meetings, New York. December, 1998.

Borrus, M., W. Sandholtz, J. Zysman, K. Conca, J. Stowsky, S. Vogel, S. Weber, 1992. *The Highest Stakes: The Economic Foundations of the Next Security System.* Oxford University Press, New York.

Brown, M., 1995. *Impacts of National Technology Programs.* Organization for Economic Cooperation and Development, Paris.

Cadot, O., H.L. Gandel, J. Story, D. Webber, 1996. *European Casebook on Industrial and Trade Policy.* Prentice Hall, New York.

Caracostas, P. and U. Muldur. 1998 *Society, The Endless Frontier: A European Vision of Research and Innovation Policies for the 21st Century.* European Commission, Brussels.

Council on Competitiveness, 1993. *Roadmap for Results: Trade Policy, Technology and American Competitiveness.* Washington, D.C.

Dasgupta, P. and P. David, 1994. "Toward a New Economics of Science," *Research Policy.* Vol. 23.

David, P., D. Mowery, W.E. Steinmueller, 1994. "Government-Industry Research Collaborations: Managing Missions in Conflict." Paper presented at CEPR/AAAS conference *University Goals, Institutional Mechanisms, and the 'Industrial Transferability' of Research.* Stanford, Calif.

European Commission, 1997. *The Commission's proposal for the Fifth Framework Programme (1998–2002).* EUR 17651.

European Commission, 1996. *The Action Plan for Innovation.* Luxembourg.

European Commission, 1995. *The White Paper on Education and Training: Toward a Learning Society.* Luxembourg.

Fields, K., 1995. *Enterprise and the State in Korea and Taiwan.* Cornell University Press, Ithaca, N.Y.

Flamm, K., 1996. *Mismanaged Trade? Strategic Policy and the Semiconductor Industry.* The Brookings Institution, Washington, D.C.

Gansler, J., 1995. *Defense Conversion: Transforming the Arsenal of Democracy.* MIT Press, Cambridge, MA.

Gaster, R. and C. Prestowitz, 1994. *Shrinking the Atlantic: Europe and the American Economy.* North Atlantic Research and the Economic Policy Institute, Washington, D.C.

Grindley, P., Mowery, D., and Silverman, B. 1994. "SEMATECH and Collaborative Research: Lessons in the Design of High-Technology Consortia." *Journal of Policy Analysis and Management.* Vol. 14 No. 4. 723–758.

Landau, R., T. Taylor, and G. Wright, 1996. *The Mosaic of Economic Growth.* Stanford University Press, Stanford, CA.

Lerner, J. 1999. "Public venture capital: Rationales and evaluation." In National Research Council, *The SBIR Program: Challenges and Opportunities.* Washington, D.C.: National Academy Press.

Ham, R., G. Linden, and M. Appleyard, 1998 "The evolving role of semiconductor consortia in the U.S. and Japan," *California Management Review*, Vol. 41, No. 1, Fall 1998. pp. 137-163.

Horrigan, J. B., 1999. *Cooperating Competitors: A Comparison of MCC and SEMATECH.* Monograph. National Research Council, Washington, D.C.

Mowery, D. C., ed., 1998. *International Collaborative Ventures in U.S. Manufacturing.* Ballinger, Cambridge, Mass.

National Research Council, 1999, *The SBIR Program: Challenges and Opportunities.* Washington, D.C.: National Academy Press.

National Research Council. 1999, *A Review of the Sandia Science and Technology Park Initiative.* Washington, D.C.: National Academy Press.

National Research Council, 1996, *Conflict and Cooperation in National Competition for High-Technology Industry.* Washington, D.C.: National Academy Press.

National Research Council, 1996. *The Unpredictable Certainty: Information Infrastructure Through 2000.* Washington, D.C.: National Academy Press.

National Research Council, 1995. *Standards, Conformity Assessment, and Trade into the 21st Century.* Washington, D.C.: National Academy Press.
National Science Foundation, 1998. *Science and Engineering Indicators.* Washington, D.C.: National Science Board.
Nelson, R.R., 1993. *National Systems of Innovation: a Comparative Study.* Oxford, Oxford University Press.
Nelson, R.R., 1996. *The Sources of Economic Growth.* Cambridge, Mass: Harvard College.
Office of Technology Assessment, 1991. *Competing Economies: America, Europe, and the Pacific Rim.* Washington, D.C.: Congress of the United States.
Office of Technology Assessment, 1995. *International Partnerships in Large Science Projects.* Washington, D.C.: Congress of the United States.
Organization for Economic Cooperation and Development, 1997. *Best Practice Policies for Small and Medium-sized Enterprises.* Paris.
Organization for Economic Cooperation and Development, 1997. *Electronic Commerce: Opportunities and Challenges for Government.* Paris.
Organization for Economic Cooperation and Development, 1996. *The Knowledge-based Economy.* Paris.
Organization for Economic Cooperation and Development, 1995. *Recommendations of the Council Concerning Principles for Facilitating International Co-operation Involving Enterprise.* Paris.
Pavitt, K., 1995. "National Policies for Technical Change: Where Are There Increasing Returns to Economic Research?" Paper presented before the Colloquium on Science, Technology, and the Economy, Irvine, CA. National Academy of Sciences.
Porter, M., 1990. *The Competitive Advantage of Nations.* Free Press, New York.
Rashish, P., ed., 1996. *Building Blocks for a Transatlantic Economic Area.* Washington, D.C.: European Institute.
Reid, Procter and Alan Schriesheim, eds., 1996. *Foreign Participation in U.S. Research and Development: Asset or Liability.* National Academy of Engineering, National Academy Press, Washington, D.C.
Rembser, J., 1995. *Intergovernmental and International Consultations/Agreements and Legal Cooperation Mechanisms in Megascience.* OECD, Paris.
Rosenbloom, R. and Spencer, W., 1996. *Engines of Innovation: U.S. Industrial Research at the End of an Era.* Harvard Business School Press, Boston, Mass.
Sanholz, W., 1992. *High-Tech Europe: The Politics of International Cooperation.* University of California Press, Los Angeles.
Saxenian, Annalee, 1994. *Regional Advantage: Culture and Competition in Silicon Valley and Route 128.* Harvard University Press, Cambridge.
Trans-Atlantic Business Dialogue: Overall Conclusions, 11 November 1995. Seville, Spain.
U.S. Department of Commerce, 1998, *The Emerging Digital Economy.* Washington, D.C.: Government Printing Office.

U.S. Department of Commerce, 1997, *A Framework for Global Electronic Commerce*. Washington, D.C.: Government Printing Office.

Wessner, Charles W. ed., 1997. *International Friction and Cooperation in High-Technology Development and Trade*. Washington, D.C.: National Academy Press.

Wolff. A., T. Howell, B. Bartlett, and R.M. Gadbaw, 1992. *Conflict Among Nations: Trade Policy in the 1990s*. Westview Press, San Francisco.

Annex

European Union Research Programs

*Professor Jorma Routti and Dr. William Cannell,
DGXII, European Commission*

International collaboration in research, involving universities, research centers, and industry, has long been supported by the European Union (EU). Organized since 1984 within successive multinational framework programmes, community research activities are designed to complement those of the EU member states and work toward closer integration of Europe's scientific and industrial communities. The central objectives of community research policy are to reinforce and mobilize the EU's scientific and technological capabilities in support of industry, the economy, and quality of life.

The Fifth Framework Programme (1998 to 2002) breaks with tradition in targeting resources on specific socioeconomic objectives, by means of focused research actions of an integrated and interdisciplinary nature. The approach will be more selective than the science and technology-driven approach of the past and will favor partnerships and networks of research actors, public and private, which are more strongly oriented toward utilization and uptake of results.

BENEFITS OF EUROPEAN COLLABORATIVE RESEARCH

Encouraging higher investment in research and technology as well as improvements in research productivity are clear economic priorities for Europe. Levels of expenditure on research and development (R&D) tend to lag those of competitors overseas. Overall, the EU spends 1.8 percent of its gross domestic product (GDP) on civil R&D, as opposed to 2.5 percent in the United States and 2.8 percent in Japan. The EU's position on patenting technological inventions is weaker than that of the United States and Japan, and Europe's major industries tend to have relatively low science intensity. Also links between industry and the university sector need reinforcement.

Action at the community level aims to promote research collaboration on a European scale that brings a number of benefits:

- Bringing together the research capabilities of research actors in different member states improves the linkages between the different types of actors (public and private) at the European level, provides a deeper pool of expertise to address existing as well as new and emerging problems, and provides a stimulus toward a more dynamic technological and business environment.
- There is an increasing number of areas of research that can only be carried out effectively on a transnational basis. Some phenomena that need to be studied are intrinsically international (e.g., climate change, marine and terrestrial ecosystems). In other areas the research effort needed surpasses the capacity of individual countries (e.g., genome sequencing).
- Large-scale research infrastructure is of crucial importance to many areas of science and technology but in view of its costs is not evenly distributed around the European Union; cross-national access can optimize its effective utilization as well as the direction of further development.

FRAMEWORK PROGRAMMES FOR EUROPEAN COLLABORATIVE RESEARCH

Nature of Framework Programme

Under the present treaties, the Framework Programme* encompasses all of the research activities carried out by the EU. It aims to strengthen the scientific and technological competitiveness of European industry and to provide support for the broad range of community policies. According to the treaty, the Framework Programme comprises four different "activities" each of which is implemented by one or more "specific programmes."

- *Research, technology development, and demonstration*, mainly through European collaborative research networks, involving enterprises, research centers, universities, and policy organizations. This activity comprises the

---

*There are in fact two Research Framework Programmes, provided for under the EC and Euratom treaties, respectively. Their content is complementary (the EC focusing on nonnuclear and the Euratom program on nuclear research), and their administration is harmonized; hence, they will be considered here under the generic title "Framework Programme."

majority of expenditures, amounting to about 87 percent of funds under the present (Fourth) Framework Programme.
- *International cooperation in research* involving partners outside the EU and/or international organizations. Such cooperation supports the development of less developed countries, provides community researchers with access to new technologies in advanced countries outside the EU, and builds research networks with neighboring countries, especially with candidates for accession to the EU.
- *Dissemination and exploitation* of research results through a variety of actions, including networks for technology transfer and innovation, support for best practice in management of research and technology, and advisory structures.
- *Stimulation of the training and mobility of researchers*, through international fellowship schemes.

Evolution of the Framework Programmes

The first Framework Programme was established in 1984 as an umbrella for a number of research activities that had been developed earlier under the European Community and Euratom treaties. Since then, yearly investments to community research have grown by a factor of three in real terms; they now amount to 3.5 billion ecu per annum. The Framework Programme accounts for 4 percent of civil government-funded research in the union. Research also represents about 4 percent of the total community budget of some 90 billion ecu (by comparison the common agriculture policy accounts for about 50 percent and the structural funds for 32 percent). When other funding arrangements, such as EUREKA for industrial research collaboration, COST for joint research funded by 25 participating countries, and those run by the European Space Agency, CERN for particle physics, the European Molecular Biology Laboratory, and the European Science Foundation are included, the total European collaborative research effort accounted for 16 percent of government expenditures on civil research in 1996, compared with 6 percent in 1985.

The majority of funding under the four Framework Programmes to date has been allocated to five broad themes: energy, life sciences, environment, industrial and materials technologies, and information and communications technologies. However, priorities have evolved over time. Energy research has diminished in relative importance; life sciences have progressively increased; and after increasing during the 1980s, information and communications technologies have declined somewhat. At the same time, a number of other research areas, such as transportation and socioeconomic research, have grown in importance, as have the horizontal activities (i.e., international cooperation, dissemination, and training and mobility).

BOX A.1
THEMATIC PROGRAMS

QUALITY OF LIFE AND MANAGEMENT OF LIVING RESOURCES

Key Actions

- *Health, food, and environmental factors*—improving health through a safe, balanced, and varied food supply for consumers covering the whole food chain and through reduction of environmental hazards.
- *Control of infectious diseases*—the fight against infectious diseases, based on new and improved vaccines, a better understanding of the immune system, and public health aspects.
- *The "cell factory"*—exploiting advances in understanding the cellular and subcellular properties of microorganisms, plants and animals, for health, environment, agriculture, chemicals, and so forth.
- *Sustainable agriculture, fisheries, and forestry, including integrated development of rural areas*—developing the knowledge and technologies needed for the production and exploitation of living resources, covering the whole production chain.
- *The aging population*—promoting the health and autonomy of older people with prevention and treatment of age-related illnesses and their social consequences.

Generic Research and Technological Development

- Chronic and degenerative diseases (particularly cancer and diabetes), cardiovascular diseases, and rare diseases.
- Research into genomes and diseases of genetic origin.
- Neurosciences.
- Public health and health services research.
- Study of problems relating to biomedical ethics and bioethics in the context of respect for fundamental research values.
- Socioeconomic aspects of life sciences and technologies within the perspective of sustainable development.

Support for Research Infrastructures

- Databases and collections of biological material, centers for clinical research and trials, facilities for fishery and aquaculture research.

CREATING A USER-FRIENDLY INFORMATION SOCIETY

Key Actions

- *Systems and services for the citizen*—fostering the creation of next-generation general-interest digital services (health, disabled, public administrations, environment, transportation) for flexible access by all citizens.
- *New methods of work and electronic commerce*—developing technologies to help companies operate and trade more efficiently and facilitating improvements in working conditions.
- *Multimedia content and tools*—future information products and services; enabling linguistic and cultural diversity; for electronic publishing, education, and training, including innovative forms of multimedia content; and tools for structuring and processing them.
- *Essential technologies and infrastructures*—promoting technologies for the information society (communications, networks, software, microelectronics, etc.), speeding up their introduction, and broadening their field of application.

Generic Research and Technological Development

- Future and emerging technologies (open-domain and proactive initiatives).

Support for Research Infrastructures

- Support for broadband interconnection of national research and education networks and advanced European test beds to assist in the development of standards, results, and applications to facilitate implementation and interoperability of advanced computer and communications systems for research.

PROMOTING COMPETITIVE AND SUSTAINABLE GROWTH

Key Actions

- *Innovative products, processes, and organization*—facilitating the development of high-quality innovative products and services and new methods of sustainable production and manufacture.
- *Sustainable mobility and intermodality*—developing integrated options for the mobility of people and goods, improving transportation efficiency, safety, and reliability, and reducing congestion and environmental disbenefits.

Continued on next page

BOX A.1 Continued

- *Land transportation and marine technologies*—developing innovative materials, technologies, and systems for sustainable and efficient land transportation, and for sustainable exploitation of the seas' potential.
- *New perspectives in aeronautics*—helping the development of aircraft, systems, and components to improve European competitiveness while ensuring rational management of air traffic.

Generic Research and Technological Development

- New materials and their production and transformation.
- New materials and production technologies in the steel industry.
- Measurements and testing.

Support for Research Infrastructures

- Support for large infrastructures through networking ("virtual institutes"), laboratories, facilities for measurements and tests and specialized databases.

PRESERVING THE ECOSYSTEM

Key Actions

- *Sustainable management and quality of water*—producing the knowledge and technologies needed for rational management of water resources for domestic, industrial, and agricultural needs.
- *Global change, climate, and biodiversity*—developing the scientific and technological understanding and tools to underpin community environmental policies and help deliver the goal of sustainable development.
- *Sustainable marine ecosystems*—promoting sustainable and integrated management of marine resources.
- *The city of tomorrow and cultural heritage*—sustainable economic development of the urban environment, improved urban planning and management, protection of quality of life and cultural identity of urban inhabitants, and restoration of social equilibria and protection of cultural heritage.
- *Cleaner energy systems, including renewables*—minimizing the environmental impacts of the production and use of energy in Europe, through research on cleaner and renewable energy sources and fossil fuel use.

- *Economic and efficient energy for a competitive Europe*—providing Europe with a reliable, clean, efficient, safe, and economical energy supply through improved efficiency and reduced costs at every stage of the energy cycle.

Generic Research and Technological Development

- The fight against major natural and technological hazards.
- Development of earth observation satellite technologies.
- Socioeconomic aspects of environmental change in the perspective of sustainable development.
- Socioeconomic aspects of energy within the perspective of sustainable development (the impact on society, the economy, and employment).

Support for Research Infrastructures

- Research installations on climate and global change, marine research, and natural risks.

EURATOM ACTIVITIES

Key Actions

- *Controlled themonuclear fusion*—the aim is to pursue the development of fusion energy as an option for clean and safe energy production; this embraces all research activities on fusion undertaken in member states.
- *Nuclear Fission*—the aim is to help ensure the safety of Europe's nuclear installations, protect workers and public, and ensure the safety and security of waste; improve industrial competitiveness; and explore new concepts.

Generic Research and Technological Development

- Radiological protection and health.
- Environmental transfer of radioactive materials.
- Industrial and medical uses and natural sources of radiation.
- Internal and external dosimetry.

Support for Research Infrastructures

- Large facilities, networks for collaboration, databases, and biological tissue banks.

traditional approach to strengthen technological capabilities and ensure access to new knowledge and expertise.
- *Research infrastructure* support optimizes the utilization and further development of infrastructure and facilities across Europe.

The horizontal programs complement the thematic programs by focusing on issues of international cooperation, SMEs, dissemination and exploitation, training, and mobility. They are common to all thematic programmes but also require specific activities.

Socioeconomic Research

In keeping with the treaty requirement to support the scientific and technological bases of European industry, the Framework Programme has been mainly concerned with natural science and technology. However, increasing importance has been given to the social and economic aspects in successive programs. This acknowledges the substantial impact of social, behavioral, and economic factors on the development and use of science and technology. It also recognizes benefits achieved from the international linkages in these areas that would otherwise be addressed in a fragmentary manner. The Fifth Framework Programme has been designed to address socioeconomic research in several fronts.

First, socioeconomic research is of importance in the thematic programs. Key actions follow an integrated interdisciplinary philosophy to optimize their economic, industrial, environmental, and social benefits. For example, in biotechnology and bioethics, transportation issues, energy and environment, and information society, socioeconomic issues are of as much concern to citizens as the science and technology.

Second, part of the horizontal program on "Improving Human Potential and the Socioeconomic Research Base" is dedicated to socioeconomic research as such, its focus being on the structural changes facing societies. Research will, for instance, be carried out on structural, demographic, and social trends; relationships between technological change, employment, and society; changing roles of European institutions, systems of governance and citizenship; and the validation of new development models.

Third, research will be promoted on science and technology policy issues and related indicators to provide a basis for the development of future policies.

The Joint Research Centre

A proportion of funding under the Framework Programme (about 7.3 percent in the Fourth Framework Programme) is allocated to the European Community's own research laboratory, the Joint Research Centre (JRC), through so-called direct

BOX A.2
HORIZONTAL PROGRAMS

CONFIRMING THE INTERNATIONAL ROLE OF COMMUNITY RESEARCH

The aims are to promote science and technology cooperation internationally, to reinforce community capacities in the fields of science and technology, to generally support the achievement of scientific excellence within the wider international framework, and to contribute to implementation of the community's external policy with the accession of new members in mind.

Actions Specific to the Horizontal Program

- *Cooperation with third countries*—activities would be differentiated by category of country: candidates for EU membership (e.g., promotion of center of excellence, facilitating participation in the other programs of the Framework Programme); NIS and other central and Eastern European countries (support for their research and technological development potential and cooperation in areas of mutual interest); Mediterranean partner countries (improving their RTD capacities and promoting innovation; cooperation in areas of mutual interest); developing countries: (sustainable management and use of natural resources, health, nutrition, and food security); emerging economy and industrialized countries (exchanges of scientists; organization of workshops; promotion of partnerships and enhanced mutual access, e.g., through science and technology cooperation agreements).
- *Training of researchers*—fellowships for young researchers from developing countries, Mediterranean, and "emerging economy" countries to work in community laboratories and vice versa.
- *Coordination*—with COST, EUREKA, and international organizations, with other external assistance activities (PHARE, TACIS, MEDA, and EDF and with member states.

INTERNATIONAL COOPERATION PURSUED THROUGH THE OTHER FRAMEWORK PROGRAMME ACTIVITIES

Participation by third countries in the specific programs may take basically two forms: (1) countries that are "fully associated" with the Framework Programme can participate on conditions similar to member states; (2) otherwise, countries may participate on a project-by-project basis (e.g., if they have a bilateral or a multilateral cooperation agreement, generally with no funding).

Continued on next page

BOX A.2 Continued

PROMOTION OF INNOVATION AND PARTICIPATION OF SMALL AND MEDIUM ENTERPRISES (SMEs)

The aim is to improve the social and economic impacts of RTD, especially the Framework Programme, through better dissemination and exploitation of research results and technology transfer, by means of policies consistent with the Innovation Action Plan, and with particular attention to the participation of SMEs in Fifth Framework Programme.

Coordination Activities on Innovation and Participation of SMEs

- *Promotion of innovation*—assuring synergy and coordination of the activities of "innovation units" to be set up in the thematic programs; definition of methods and mechanisms to improve the exploitation of results.
- *Encouraging SME participation*—support for SME participation in RTD and demonstration activities to be carried out in the programs, including cooperative research activities and exploratory awards.

Actions Specific to the Horizontal Program

- *Promotion of innovation*—activities to improve the level of uptake of technologies and results; new approaches to technology transfer; integrating the technological, economic, and social aspects of innovation; coordination of studies and analyses on innovation policy.
- *Encouraging SME participation*—a special entry point for SMEs, providing help and assistance on research programs; common instruments to harmonize and simplify SME access; economic intelligence to help SMEs identify and meet their current and future technological needs.
- *Joint actions innovation/SMEs*—rationalization, coordination, and management of networks for promoting research and innovation, electronic and other information services; providing information and assistance on the community's research and innovation activities; providing information and pilot activities on intellectual property rights; access to private finance; and assistance for the creation and development of innovative start-ups.

IMPROVING HUMAN POTENTIAL AND THE SOCIOECONOMIC RESEARCH BASE

The aim is to preserve and help develop the community's knowledge potential through greater support for the training and mobility of researchers, by enhancing access to research infrastructures and making Europe attractive for research investment; to mobilize research on the social and economic sciences and humanities to understand critical economic and social trends and requirements; and to support the community's science and technology policies.

Actions Specific to the Horizontal Program

- *Support for training and mobility of researchers*—research training networks focusing on young researchers at predoctoral and postdoctoral levels; a system of Marie Curie fellowships, including fellowships for young high-quality researchers; fellowships awarded to young researchers and hosted by enterprises (including SMEs); fellowships in the less favored regions of the community; fellowships for experienced researchers to promote mobility between industry and academia; and support for short stays by doctoral students in training sites.
- *Enhancing access to research infrastructures*—enhancing international access to research infrastructures; networks of cooperation between infrastructures; RTD projects oriented toward infrastructure.
- *Promoting scientific and technological excellence*—stimulating the exchange of scientific and technological excellence and making the most of research achievements (e.g., through high-level scientific conferences, prizes for high-quality research, actions to improve understanding of science and technology).

Key Action

- *Improving the socioeconomic knowledge base*—improved understanding of structural changes in Europe to better manage them and help citizens build their future; social trends and structural changes; technology and society; governance and citizenship; new models of development favoring growth and employment. Defining the knowledge base for employment-generating social, economic, and cultural development and for building a European knowledge society.
- *Support for the development of science and technology policies*—strategic analysis of key policy questions; development of a common base of science, technology, and innovation indicators; supporting the development of the specific knowledge base needed by policymakers and other users on European science and technology policy issues.

Action Pursued through Other Framework Programme Activities

The horizontal program would provide coordination, support, and accompanying actions needed to ensure consistency with action undertaken elsewhere in the Framework Programme on aspects related to the objectives and activities of this program.

actions. The JRC's main role is to provide neutral and impartial scientific and technical support to the development of community policies and regulations.

The activities of the JRC are focused on areas where its skills and equipment—in many cases unique to Europe—can provide added value, through client/supplier relationships with the commission's policy work. JRC is participating also in "indirect" actions under the Framework Programme, as partner in trans-European consortia, where it competes with other research proposers in the normal manner.

CONCLUSIONS

This paper describes the role and broad objectives of the Framework Programme and gives a picture of the program in the coming years. The program is both a political instrument, designed to deliver tangible results for community policies and innovation, and a funding mechanism sensitive to the demands of national participants. The program has benefited European research but nevertheless needs to be updated and made more strategic. By focusing effort at the community level on truly strategic challenges for Europe, it is hoped that the program will have a more profound impact to the benefit of the European Union's citizens.

REFERENCES

Fifth Framework Programme for Research and Technological Development (1998-2002), Commission Working Paper on Specific Programmes: Starting Points for Discussion, Brussels.

European Commission, Science, Research, Development: The Commission's Proposal for the 5th Framework Programme (1998-2002).

European Commission, Science, Research, Development: Towards the 5th Framework Programme – Scientific and Technological Objectives.

European Commission, Science, Research, Development: Inventing Tomorrow, Europe's Research at the Service of Its People, Preliminary Guidelines for the Fifth Framework Programme.

Green Paper on Innovation, Bulletin of the European Union, Supplement 5/95.

Multilingual Information Management

Gary Strong
National Science Foundation

MOTIVATION

Every day new computer users link to the World Wide Web and more information is made available on it. Most of this growth is in Europe and Asia; by one estimate, annual growth for English documents is 50 percent and for other languages over 90 percent.

Nobody can speak all of the world's languages. Therefore, every individual, every business, and every government suffers a strategic gap in the new world infostructure. Technology must be developed to support the locating, translating, browsing, and dissemination of multilingual information of both spoken and written language. This technology can be deployed in many ways. Compelling applications include:

- Commerce—technology watch for foreign business opportunities and challenges, tourism and commerce via the World Wide Web and e-mail;
- Education—learning and browsing from foreign sources (for term papers, projects, etc.) and foreign-language instruction, even just to browsing level;
- Government—information dissemination to citizens in various languages (of tax forms, census information, etc.) and information assimilation for strategic purposes (environmental monitoring, intelligence, etc.).

CURRENT SITUATION

Despite considerable advances in language-processing technology required to support multilingual information access in both the United States and Europe, what is not yet available, not even pilot research systems, is an integrated multilingual information access service that includes voice and typed modalities, IR, MT, and browsing and display capabilities. The core technical obstacles that must be addressed fall into three classes:

- *Cross-lingual IR and MT*—This entire research area is so young that even the broad outlines of how to proceed are controversial. Basic research is urgently needed to determine which of these directions would be the most fruitful to pursue. Example directions are investigating various combinations of IR and MT, including how exactly IR and MT should be interrelated; identifying the lexical and semantic asymmetries between languages in order to develop methods to avoid cross-language error proliferation; and investigating novel techniques that do not combine IR and MT, such as translingual Generalized Vector Space Models or Latent Semantic Indexing.
- *Multilingual speech processing*—Research is required in the core area of multilingual speech recognition and synthesis, including the deployment of language identification and appropriate language switching, in an integrated system.
- *Large multilingual digital information collections*—Research is required in the collection, standardization, deployment, and maintenance of text resources from the government (including the Census Bureau and Library of Congress) and business sectors (including encyclopedias and technical manuals).

Three support efforts are required to ensure success:

- *Resources*—Building on the European Union's standardization experience in the EAGLES project, lexical, textual, semantic, and spoken resources and corpora must be collected, standardized, annotated, and supported to enable rapid and coordinated research and development (R&D) toward robust processing and wide coverage.
- *Evaluation*—Building on the U.S. experiences with TREC (for IR), MT EVAL (for MT), MUC (for information extractions), several speech recognition evaluation programs, and others, an evaluation paradigm must be designed for multilingual access tasks, and a series of evaluation meetings must be held to measure the progress of research.
- *Interfaces*—The interfaces currently used by IR, MT, speech, and related systems were not designed to support interactions with integrated multilingual systems. Appropriate designs are required to handle, in a fully integrated way, speech recognition, IR, MT, document display and browsing, speech synthesis, summary generation, and so forth.

RECOMMENDATIONS

Participants at the First International Language Resources and Evaluation Conference (LREC) and the workshop sponsored by the National Science Foundation on multilingual information management identified the need for research and development in multilingual information management and access that would coordinate and integrate efforts languages as well as applications and core technologies (e.g., syntactic analysis, sense disambiguation). To this end they emphasized the need to establish an international collaborative framework to support and define the effort.

By the nature of multilingual information access and with the current capabilities of various research groups in the United States and European Union, a collaboration between the two regions would benefit both. The mutual complementarity exists because of the following:

- *U.S. strengths*: (1) Breadth and depth of all aspects of English IR, MT, speech processing summarization, and core technology (the United States has spent as much on R&D in English, as the European Union has spent on all its various languages combined); (2) significant competence in several Asian languages, including Japanese, Chinese, Arabic, and Korean, as well as access to students from the Far East and India; and (3) experience in multilingual language processing evaluation, including problems of TREC and MUC.
- *EU strengths*: (1) Well-established efforts to create standards for linguistic resources (in particular, EAGLES and SIMPLE); (2) experience with resource creation, including multilingual resource creation, that implements these standards; (3) experience with cross-lingual issues in integrated multilingual systems; and (4) breadth and depth in European languages, notably in languages of non-Indo-European origin such as Finnish, Basque, and Turkish, as well as access to languages of Eastern Europe, the Middle East, and China via several contracts.

The participants articulated key areas to be addressed by this collaborative effort in support of multilingual information access: annotated, standardized multilingual resources; tools to develop the resources; evaluation paradigms to assess the state of the art and encourage progress in various technologies. These concerns reflect the growing importance and demonstrated success of data resources to support automated language processing applications. The resources range from balanced text and speech corpora to corpora annotated for the occurrence of linguistic phenomena (names, noun phrases) to higher-level resources, such as lexicons with syntactic and semantic information. Resource development represents an excellent area for international cooperation, since each country has access to materials in its language, the linguistics expertise to provide reliable annotation, and the motivation to provide its citizens and industry with monolingual information access tools for that particular language, as well as multilingual access to the global information space.

Multilingual Resource Creation

Work in multilingual information processing requires assembling balanced comparable or parallel corpora for multiple languages, together with linguistic resources such as translation lexicons, including syntactic and semantic information, ontologies (WordNet, EuroWordNet, SIMPLE), and so forth. In addition to supporting the creation of resources in itself, one of the key challenges is to develop standardized data formats and annotation paradigms that are applicable across languages and applications.

Europe has already invested considerable effort in both resource creation and standards development (via projects such as EAGLES and PAROLE). Because of the multilingual environment, Europe has also considered lexical and semantic asymmetries between languages, which is required to develop core technologies applicable across languages and avoid cross-language error proliferation. U.S.-European cooperation would enable the United States to benefit from the European experience in these areas; as a start, the National Science Foundation workshop participants recommended that the United States join the EAGLES standardization effort as soon as possible. It is important to point out that linguistic resources must be made freely (i.e., without legal incumbrance) for research use.

Tools

One major challenge is to provide tools that enable multilingual resource creation and annotation. Such tools would dramatically decrease the cost of resource creation, standardize data formats, and improve the quality of data annotation. These tools must apply across multiple languages, multiple writing systems, and multiple media (speech, language, and video). There is already significant international cooperation—namely the Transcriber tool (collaboration between DGA/DCE/CTA/GIP France, and the LDC, United States), and collaboration on a multilingual text processing environment, GATE, between the University of Sheffield in the United Kingdom and NMSU in the United States. If we add to this a freely available multilingual annotation tool (e.g., Mitre's Alembic Workbench), together with a syntax annotation tool (e.g., DFKI's syntax annotation tool, the SIMPLE syntax and semantic tools), we would obtain a powerful, uniform corpus preparation environment. Tools for the creation of other resources, such as lexicons, also exist, which could potentially be adapted to a common environment.

Evaluation

The technology evaluation paradigm of the Defense Advanced Research Project Agency has proved remarkably successful in fostering progress in key areas, including speech understanding, information extraction, and information retrieval. For speech this evaluation paradigm has resulted in a decrease in word error by a factor of two every two years and commercially available medium-to-large vocabulary recognition systems. A number of EU/U.K. groups (Cambridge, Phillips, LIMSI) have participated with great success in these evaluations, in addition to both university and industry groups in the United States (e.g., Dragon, IBM, CMU, SRI, OGI).

In information extraction the introduction of component metrics for name identification has resulted in rapid commercialization of multilingual name extraction technology. In addition, there are now comparable kinds of evaluation going for French under the auspices of Aupelf-Uref. The text retrieval workshop has been international from its inception. Thus, evaluation is another obvious place to combine U.S. expertise in evaluation and EU expertise in multilingual corpus creation and standardization to increase cooperative multilingual evaluations in key technology areas, for example, broadcast news understanding, cross-language document retrieval, and word sense disambiguation.

SUMMARY

A research program of approximately five years is recommended to address the core obstacles preventing a fuller understanding of multilingual information access and to begin to foster a suitable climate for commercial exploitation. Such a program should include the following broad thrusts: development and support of multilingual resource creation, standardization, and maintenance; development of an evaluation paradigm and support of and evaluation program; and investigation of tie-ins of multilingual information access to related IT areas.

We recommend that this be accomplished by establishing an international U.S.-EU collaborative effort, with two major aims: provision of a broad framework for research that specifically addresses issues and technologies across languages, applications, etc., and provides for interaction, feedback, and collaboration across these areas, and joint development of technology through the fostering of collaborative projects between Europe and the United States. The outcomes of such a program should, if successful, ultimately enable application in commerce (e.g., technology watch), education (e.g., report writing), and government (e.g., environment monitoring) across a wide variety of languages and domains.

Charge for Electronic Commerce Subgroup

Ray Kammer, Director,
National Institutes of Standards and Technology

The Organization for Economic Cooperation and Development and other international organizations have expressed their belief that "the exponential growth and diffusion of the Internet [are] quickly making the promise of widespread electronic commerce a reality. High-speed, interconnected global networks like the Internet provide new ways to conduct commercial transactions, generate new markets and revenue streams, lower transactional costs, and forge new relationships between businesses and consumers."

The goal of this subgroup is to identify concrete opportunities for cooperative activities between the United States and the European Union in the area of electronic commerce. We will seek to identify gaps in current research plans and work together to establish joint projects to fill them. Common issues and questions will be noted for discussion at a subsequent conference.

As an example of some potential areas of discussion, attendees of a recent public industry workshop on electronic commerce in the United States delivered the following major conclusions:

- A lack of interoperability among technologies is a major inhibitor to electronic commerce.
- A window of opportunity exists to put a framework in place before the market fragments into incompatible point solutions, thereby slowing the evolution of electronic commerce.
- Access to trusted information through electronic commerce will empower the consumer and businesses of all sizes.

The attendees further identified the following areas of innovative technology as being promising possibilities for future research and development: multi-application smart card infrastructure, distributed intelligent search protocols, intelligent markets, webs of trust, and public key infrastructure (PKI).

We are very fortunate to have here a distinguished group of experts representing industry and government from both sides of the Atlantic Ocean. Each speaker will present a brief presentation, which will be followed by questions and answers and discussions. I am pleased to introduce the first speaker—Ric Jackson, from the National Institute of Standards and Technology.

SAMPLE QUESTIONS TO CONSIDER DURING THE DISCUSSIONS

From OECD, the Organization for Economic Cooperation and Development, "Dismantling the Barriers to Global Electronic Commerce," November 1997, Turku, Finland: (1) How best can technologies and policies be developed to help protect or aid in the prosecution of intellectual property rights violators? (2) How can technological solutions be used to protect consumers? What are the best mechanisms for developing and deploying these solutions? (3) To what degree can technological solutions give users confidence that their privacy is being protected? What complementary solutions are needed? (4) To what degree can technological solutions instill trust in electronic commerce? (5) What is the best mechanism for making commercial codes compatible with global electronic commerce?

Today, information technology influences almost all areas of human endeavor, creating new products, services, and industries. Vice President Gore said that "we are on the verge of a revolution that is just as profound as the change in the economy that came with the industrial revolution. Soon electronic networks will allow people to transcend the barriers of time and distance and take advantage of global markets and business opportunities not even imaginable today, opening up a new world of economic possibility and progress."

It is my pleasure to have been invited to chair the breakout session on information technologies. As you have already heard, the U.S. government and the European Commission recently signed a new science and technology agreement to enhance U.S.-European cooperation on a broad range of science and technology issues.

Information technology is one of the important issues that has been selected for discussion at this session. I would like to introduce the U.S. and the European representatives who have been instrumental in organizing this discussion session: EU Representative—Thierry van der Pyl, ESPRIT, and U.S. Representative—Tom Kalil, National Economic Council, White House.

The breakout sessions are designed to identify concrete opportunities for cooperative activities in different thematic areas by bringing together policymakers and researchers from both sides of the Atlantic. The information technologies session will split into three subgroups. Each subgroup session will include researchers from government, universities, and industry, as well as policymakers from both sides of the Atlantic. The topics for the subgroups are next-generation Internet, electronic commerce, and translingual information management.

We are very fortunate to have participation from top-level officials and an exceptionally broad range of experts. I thank Tom Kalil for agreeing to chair the are next-generation Internet subgroup and Gary Strong and Roberto Cencioni for chairing the translingual information management subgroup. I have the honor of chairing the e-commerce subgroup.

Common issues and questions will be noted for discussion at a subsequent conference to be held in Europe in the fall or to be taken up in official channels such as the Joint Consultative Group.

White Papers on Transportation Research

U.S. Department of Transportation

OPPORTUNITIES FOR EU AND U.S. COOPERATION IN GLOBAL NAVIGATION AND APPLICATIONS

The Global Navigation Satellite System (GNSS) is being defined by the International Civil Aviation Organization for use worldwide. This global resource—currently based on the U.S. Global Positioning System (GPS) and the Russian GLONASS system—will be the first to provide the entire world with a single-position, navigation, and timing system that will be equally accurate and accessible in all populated areas of the world. The GNSS system will be able to be used by all modes of transportation and by many other users (e.g., surveying, geodesy, farming, telecommunications firms, power distribution, international timing, weather prediction). Currently, through the auspices of many international organizations, the European Union and the United States are already sharing information and research in the areas listed above. However, the European Union and the United States have the opportunity to cooperate in several new areas.

Spectrum Protection

- *Allocation protection.* Signals that are used by both GPS and GLONASS are currently protected internationally via their allocation status of Aeronautical Radionavigation Service. Users in the adjacent bands would like to change the rules to allow potentially interfering signals to be transmitted in this same band. This infringement could cause the entire GNSS program to be seriously hindered in its development. The United States would like to discuss ways to prevent this encroachment, ways to deal with other interfering signals in the band, and technologies to monitor the band to ensure compliance.

- *Security.* The GNSS system will allow for improvements in many areas of critical infrastructure. As such it will become the backbone for both critical safety and economically critical systems. This makes the GNSS system both a tool and a target for military users, both friendly and hostile, and other hostile users such as terrorists.
- *Military testing notification system.* Owing to its ability to provide precise positioning and navigation, the GNSS system is a candidate for exploitation by hostile military forces. For this reason, friendly nations' militaries will develop means of denial for the signals. Testing of the denial equipment and the training regimen for the friendly forces could cause harmful interference to "safety-of-life" systems. A system of data sharing and transmission of alerts for both domestic and international travelers needs to be developed that will prevent accidents from occurring because of military testing and training. This system can be a new system developed just for the GNSS system or an evolution of an existing alert network such as the ones used for distributing Notice to Airmen and Notice to Mariners. European Union participation in defining the new or evolved system is necessary for a uniform and seamless worldwide system to come to fruition.
- *Civil signal protection from interference.* The GNSS signals are very low-power signals and as such can be easily denied because of interference or overwhelmed by false signals transmitted by hostiles. The United States is interested in defining civilian means of dealing with these threats to the GNSS system given its safety and economic importance. The response to the threat can come in several levels that deal with nonharmful interference, unintentional interference, and harmful intentional interference. The United States would like to work with the European Union to expand its current security ties for early identification of parties that may try to intentionally cause harmful interference for safety or economic terrorist reasons. In addition, we need to work together to develop technologies to identify, eliminate, or respond in some fashion to all other kinds of interference that originate from unexpected sources and the means for sharing data on these types of occurrences.

Definition of the Next-Generation GPS

- *Third civil signal.* In March 1998, the Vice President announced that the United States would be modifying a military signal for use as a second civil signal and adding a third civil signal to the GPS satellites. One of these signals will be chosen to be the second "safety-of-life" signal and will require comparable international allocation protection as we are pursuing for the current civil signal. The third signal will be available for use by the scientific, precise positioning, and other users (such as weather forecasters, surveyors, and telecommunications firms) that may be able to tolerate short-duration outages caused by localized interference. The United States would like to have input from the European Union for selecting the second "safety-of-life" signal and participation in the definition of the new signal structure for the third civil signal.

- *Constellation evolution/augmentation.* The current GPS constellation is undergoing a review by the U.S. government. The future constellation may look different from the current constellation. In addition, the GPS constellation will be supplemented by other spacecraft to fulfill an augmentation role in the GNSS or for other purposes. Involvement in the definition of the look of the future constellation and the signals from these other satellites would be beneficial to both the United Stats and European Union.

Sharing of Operational Data

- *GNSS database of status reports.* To provide GNSS users with the most up-to-date information, a means of sharing health and status reports must be developed. These reports allow users to know what satellites are not healthy and why and what augmentation systems are having difficulties and why. Armed with this knowledge, users will be able to make informed choices that affect them. For example, aircraft will be able to make dynamic route-planning changes, and surveyors will be able to more efficiently schedule when and where they will be able to work. Cooperation with the European Union to develop the database standards and interfaces for this type of system is essential to the success of GNSS.

INTELLIGENT TRANSPORTATION SYSTEMS: SURFACE APPLICATIONS

The United States and the European Union have both recognized the potential for using intelligent transportation systems (ITSs) to improve the mobility, safety, and productivity of their transportation systems. Since the early 1990s both groups have implemented aggressive programs of research, operational testing, and deployment support (architecture, standards, training, etc.) for ITSs. There have been three formal exchanges between the ITS staff of the U.S. Department of Transportation (DOT) and the staff of the European Commission in DGXIII, DGVII, most recently at the ITS World Congress in Berlin. In addition, there have been informal exchanges of technical and programmatic information through the World Road Association and the annual world congresses. The exchanges to date have largely been at a high level, without sufficient follow-up or staff support. As ITS in the United States and Europe moves from an era of research to one of deployment, the importance and value of the exchange of information and technology will increase. Significant areas for cooperation are discussed below.

Year 2000 Problem

Many ITS legacy systems were programmed to use just two digits to keep track of the date. On January 1, 2000, these systems could recognize a "double zero" not as 2000 but as 1900. They could stop running or start generating inaccurate data. Among those potentially at risk are computers that operate ITS systems, such as synchronized traffic signals, electronic tolls, and automatic vehicle locators used by buses or trucks. Work is under way at DOT to determine the impact of the year 2000 computer problem on ITSs and to identify solutions. A national summit, hosted by DOT, will bring together state and local transportation officials, business leaders, transportation technology suppliers, and others to evaluate the year 2000 problem's possible effects on ITSs, identify solutions, and develop ways to promote their implementation nationwide.

Since the ITS legacy systems in place in Europe are similar to those in the United States, it is expected that the year 2000 problem is an important European issue as well. The extent of the problem in Europe and the solutions that have been identified are not well known in the United States. Collaboration on this issue can prove beneficial for minimizing the international impacts of the year 2000 problem on mobility, safety, and productivity.

Standards

Both the United States and the European Union have been aggressively pursuing the development of standards. The United States has been working through standards development organizations (e.g., Institute of Transportation Engineers, American Association of State Highway and Transportation Officials, Institute of Electronics and Electrical Engineers) to develop consensus standards. The Europeans have been working through the European standardization organization, CEN.

Both groups will be faced with the issue of implementing the standards in their respective countries. Discussions on alternate approaches for encouraging the use of standards and the role of government in testing and certifying standards may prove beneficial. There is also the need to continue discussions on harmonization of standards between the United States and Europe. There has been some discussion through the International Standards Organization. All parties agree that there is value in being more collaborative in the development of standards;

however, not all ITS standards require international harmonization. We should target only those few where there is significant payoff.

Architecture

The United States has pursued a national top-down approach through the development of a national ITS architecture using an aggressive consensus-building effort throughout the development. A total of 30 regional public fora designed to obtain stakeholder feedback for guiding the national ITS architecture were held at several stages of development. The national architecture was also approved by ITS America in 1996. DOT is now maintaining the architecture, conducting training, providing deployment guidance, and promoting architecture consistency.

The Europeans have taken a much more bottom-up approach, focusing on a more carefully designed research-oriented effort. The European Commission's Transport Telematics Applications Programme (T-TAP) has been working to consolidate system architecture development efforts through 64 projects covering all transportation modes. In addition, the System Architecture and Traffic Control Integration Task Force, initiated in 1994, has developed a recommended methodology for developing a Pan-European ITS architecture.

There may be some benefit from selective harmonization between the two efforts. ITS America has formally requested that the ITS Joint Program Office launch an exchange between the United States and Europe for the purpose of tying to harmonize architectures wherever possible. There are certainly opportunities to share lessons learned as both groups move forward using ITS architecture as a tool for facilitating the planning, design, and implementation of integrated systems.

Intelligent Vehicles

Both the United States and Europe have been pursuing research on advanced vehicle control and safety systems (AVCSSs). Recently, the United States has consolidated its vehicle-based efforts under the Intelligent Vehicle Initiative (IVI). This effort combines ongoing research on crash avoidance and automated highway systems into a single program focused primarily on improving highway safety. The IVI will use an evolutionary approach, working in partnership with industry to develop technologies that enhance driver performance.

The Europeans have also been working to improve highway safety through the development of AVCSSs. The Program for a European Traffic with Highest Efficiency and Unprecedented Safety, Dedicated Road Infrastructure for Vehicle Safety in Europe II, and T-TAP efforts have included significant research in the AVCSS area. These efforts have included strong industry participation and both autonomous (vehicle-based) and cooperative (vehicle and infrastructure-based) systems.

Significant opportunities for cooperation, including joint research, exist in the intelligent vehicle arena. To date, these have not been pursued.

Deployment

The United States has had significant technological accomplishments through its ITS research and operational testing efforts. ITS is now beginning to be deployed across the country. A primary concern is the degree to which the systems being deployed are being integrated across agency, jurisdictional, and modal boundaries. Transportation system planning and implementation processes must be changed to overcome this challenge. The federal ITS program includes a multipronged approach of model deployment, technical guidance, training, and funding incentives to facilitate integrated deployment.

The European Union faces similar, perhaps more difficult, problems in trying to achieve integrated deployment throughout Europe. Individual countries have formed organizations to facilitate deployment through cooperation between government and industry. The European Union and the European Road Transport Telematics Implementation Coordination Organization (ERTICO) provide umbrella organizations to develop overall implementation strategies and cooperation. However, the follow-through on ITS deployment is largely left up to the individual countries, with varying approaches on issues such as architecture, public versus private-sector roles, and so forth.

Given that both parties are moving much more heavily into the deployment of ITS, that there is value to integrated deployment, and that there are clear barriers to achieving this integration, there are significant opportunities to exchange information on strategies and approaches.

INTELLIGENT TRANSPORTATION SYSTEMS: MARITIME SAFETY

The U.S. Coast Guard has a statutory responsibility to ensure the safety and environmental protection of U.S. ports and waterways. In the course of meeting that responsibility the Coast Guard oversees the Vessel Traffic Service (VTS) program. A National Dialogue Group composed of national representatives of maritime, port, and public stakeholders was convened to define the basic elements of a VTS and to identify mariners' information needs. This group concluded that a VTS is an important tool for ensuring safety in a waterway. The National Dialogue Group specifically recommended an emerging technology, Automatic Identification Systems (AIS), as the principal technology for new VTS systems and to improve navigation safety in non-VTS areas.

The AIS Concept

An AIS uses radio transponders carried on vessels. An AIS transponder repeatedly broadcasts vital information about the vessel. This information may include important data about the vessel such as name, type, position (using differential GPS integrated into the system), course, speed, navigation status, dimensions, or type of cargo.

When coupled with an appropriate display capability, the AIS transponder approach gives real-time navigation and vessel traffic information to the mariner in the wheelhouse. An important aspect of AIS is that it electronically exchanges digital information between all AIS-equipped vessels and thereby reduces the intrusive voice radio traffic associated with congested ports.

Within a VTS area, AIS transponders will manage the exchange of data between vessels and the shore-based Vessel Traffic Center (VTC). The VTC would receive signals from every transponder in range, combine them, and then retransmit necessary data to all participating vessels in the VTS area. The VTS could enhance the information by including up-to-the-minute water depths, weather, current speed and direction, or other safety-related information. The mariner may then consult the display to make better decisions on collision avoidance and navigation.

Outside a VTS area, AIS transponders would work in ship-to-ship mode. The transponders would broadcast information and would in turn receive information from transponder-equipped ships nearby.

International Implications

In order for AIS to be universally adopted, it requires international cooperation in developing functional and technical standards. It will further require the universal adoption of highly accurate navigation and positioning and agreement on communications protocols. International standards for shipboard displays and data exchange also will be needed. The Coast Guard is pursuing further development of international standards as a high priority.

STRATEGIC ENABLING RESEARCH

Research is derived from our need to respond to a changing world. Safety and security concerns, land-use and demographic trends, globalization of trade and economic growth, environmental preservation, and social policy concerns are some key challenges driving transportation research, in particular. Enabling research provides a foundation for making steady technological advances and fostering breakthroughs that will be required to meet twenty-first century transportation needs. This is done by creating new transportation-related technologies and building from other related systems and strategies. Six focus areas are identified in a recent U.S. Federal Transportation Science and Technology Strategy as those that will foster innovative and cutting-edge research for developing future transportation technologies to help address these challenges. Since enabling research will have a longer-term focus and be of higher risk, it will tend to have broad applications to multiple aspects of the transportation system (vehicles, infrastructure, and human performance). The six areas and opportunities for collaboration (in italics) are as follow:

- *Human performance and behavior* (e.g., simulation, adaptive automation, fatigue monitoring, information fusion). Collaborative research in this area could be aimed at understanding how transportation system users and operators perceive, process, and act on information in real-world situations.

Two U.S. multimodal research initiatives on human factors have been identified to guide development of technologies and procedures that maximize human safety and efficiency in transportation-related activities—one explores *advanced institutional technology* and the other *alertness and fatigue.*

- *Advanced materials and structures* (e.g., fiber-reinforced plastics, new steel alloys, composite materials, and adhesives). Recent technical advances have produced a wide variety of new materials and techniques to support research on new generations of vehicle components, vehicle propulsion systems, and transportation-related construction materials and techniques.
- *Collaboration on the application of materials advances* to the transportation infrastructure could include demonstrations of their effectiveness, long-term viability, and cost competitiveness in enhancing safety and performance.
- *Computer, information, and communications systems* (e.g., software assurance, high-confidence systems, modeling, simulation, networks/next-generation Internet, wireless communications). Modern transportation systems require accurate and timely information as innovation through information infrastructures is integrated into transportation system elements and functions. To improve the efficiency, safety, and performance of these innovations, research and technology development needs to focus on system concepts and on the characterization of alternative configurations and technical choices.

High benefits are expected from *information and software assurance research* to prevent against cyber-attack.

- *Energy, propulsion, and environmental engineering* (e.g., fuel cells, energy conversion, and storage). Numerous U.S. federal government research projects seek to reduce the environmental impacts of transportation vehicles, operations, and systems. Among these projects are efforts to develop and test new energy storage and vehicle propulsion systems such as fuel cells, which produce electrical energy from fuel without combustion, and flywheel batteries, which store kinetic energy directly. New energy storage and vehicle propulsion systems like these offer enormous potential benefits for energy efficiency and emissions reductions and may be applicable to several modes of transportation.

- *Fuel cells; batteries; and hydrogen production, distribution, and storage* are expected high-payoff research focus areas for fuel-efficient, environmentally benign vehicles.
  - *Sensing and measurement* (e.g., chemical/biological hazard detection, environmental monitoring, nonobtrusive structural testing and repair, nano/micro-sensors and devices). Research in this area could be focused on development and application of technologies to monitor, analyze, quantify, and thus improve the safety and performance of transportation systems. "Smart" structures, such as roads and bridges embedded with sensors, have the potential to increase safety by providing real-time information on travel conditions. Similarly, "smart" vehicles may improve their performance by sensing environmental and operating conditions.

Nanotechnologies for continuous monitoring of human and system behavior and performance are expected to be beneficial in numerous waysæimproved safety, reduced traffic management and travel times, improved incident management and responsiveness, and increased throughput of existing physical infrastructure.

- *Analysis, modeling, design, and construction tools.* Research in this area could focus on developing information and techniques to evaluate system design improvements and to estimate the performance benefits of innovations on management of system operations. Specific collaborative efforts could be focused to develop transportation system design tools and methods to support (1) broad system engineering and integration to assure high-level system performance; (2) system performance and impact characterization to monitor and forecast the effectiveness of congestion relief and mobility enhancement strategies; (3) transportation and logistic system operations and management to assess the safety implications of planning and design decisions; and (4) transportation planning, economics, and institutions to evaluate multimodal tradeoffs for optimizing transportation expenditures among various modes.

Tools for policy research on transportation issues, such as global climate change, land use, pricing, and societal concerns, are expected to return high benefits.

INTERMODAL TRANSPORTATION:
INTERMODAL TRANSPORTATION NETWORKS

Efficient intermodal transportation networks are particularly vital to Europe and North America. Increasing volumes of goods and passenger traffic, along with growing demands for speed, safety, and environmental protection, have increased the need for interconnected transportation networks on an international scale. To achieve a balanced intermodal network, both regions recognize the necessity of integrating long-range planning priorities; the potential of technological developments; and the organizational, legal, and institutional shifts needed to improve interconnection and interoperability. There must also be greater attention to the information common to all modes and countries and how this information can be harmonized into a standard packet of information. Significant benefits from information exchange and technology cooperation have been identified for the following areas:

Legal and Regulatory Issues

Intermodalism has demonstrated advantages in reducing congestion and offering competitive modal pricing and choices in the United States and in the European Union (EU) member nations. The EU is still in the process of implementing consistent regulatory issues across its member states, particularly regarding to the liberalization of European railways. The EU's Directive 91/440 on rail deregulation addressed many problems targeted by the U.S. Staggers Rail Act. The United States can offer experience in regulatory oversight gained from recent rail mergers and acquisitions.

The United States and the EU face similar challenges in rail consolidations, mergers, and service decisions that must be acted on to achieve greater transportation efficiency but ensure that customers and consumers are not disenfranchised.

Physical Infrastructure Constraints

EU transportation officials recognize that the majority of physical transportation impediments involve access constraints—incompatible land use, roadway congestion, rail service limitations, poor timing/coordination of infrastructure improvement projects—that hinder the movement of cargo into and out of intermodal terminals and points of manufacture/distribution. By fostering the development of trans-European networks, the European Commission has made a substantial effort to stimulate the growth of system-wide intermodal movements as a genuine alternative to road transportation. The United States is taking steps to eliminate transportation bottlenecks through requirements for coordinated transportation planning, assessing and investing intermodal connections to the

National Highway System, and identifying long-term strategic needs through the Waterways Management Initiative.

Equipment Standardization

European intermodal stakeholders have identified the manufacture, use, and regulation of equipment, standardization of containers and rail equipment, and the use of the swap body (European truck trailers) as issues requiring further discussion.

The United States and EU share common problems trying to utilize domestic equipment in international transportation. The United States has initiated discussions on standardization with Mexico and Canada in accordance with the North American Free Trade Agreement.

Information Processing

Data collection to support regulatory safeguards and trade analysis will be a major concern to the European Union when it eliminates all border crossings and attendant information checks.

Data processing and collection constitute a major enabling factor for intermodal transportation growth in Europe; this electronic interchange includes four components: (1) the need to coordinate actions of both small and very large organizations active in intermodal transport chains; (2) the need to involve economic actors from different modes, many of which have very divergent views and concepts of how to respond to customer needs; (3) the presence of a very wide divergence in technology applications introduced by the various parties engaged in intermodal transportation; and (4) the international character of many intermodal chains.

The United States has embarked on an effort to facilitate and expedite the collection of international trade data through the creation of an International Trade Data Systems (ITDS) initiative; the Office of Intermodalism represents U.S. Department of Transportation on the ITDS Board of Directors.

Agreement for Scientific and Technological Cooperation Between the European Community and the Goverment of the United States of America

The following Agreement was signed on December 5, 1997 in Washington, DC.

THE EUROPEAN COMMUNITY (thereinafter "the Community"), of the one part, and

THE GOVERNMENT OF THE UNITED STATES OF AMERICA, of the other part, hereinafter referred to as the "Parties";

CONSIDERING the importance of science and technology for their economic and social development;

RECOGNIZING that the Community and the Government of the United States of America are pursuing research and technological activities in a number of areas of common interest, and that participation in each other's research and development activities on a basis of reciprocity will provide mutual benefits;

HAVING REGARD to the Declaration on EC-US Relations of November 23, 1990, and the New Transatlantic Agenda and the Joint KU-US Action Plan adopted in Madrid on December 3, 1995;

DESIRING to establish a formal basis for cooperation in scientific and technological research which will extend and strengthen the conduct of cooperative activities in areas of common interest and encourage the application of the results of such cooperation to their economic and social benefit;

HAVE AGREED AS FOLLOWS

ARTICLE 1

Purpose

The Parties shall encourage, develop and facilitate cooperative activities in fields of common interest where they are pursuing research and development activities in science and technology.

ARTICLE 2

Definitions

For the purposes of this Agreement:

(a) "Cooperative activity" means any activity which the Parties undertake, or support, pursuant to this Agreement, and includes joint research;

(b) "Information" means scientific or technical data, results or methods of research and development stemming from joint research, and any other data relating to cooperative activities;

(c) "Intellectual Property" shall have the meaning defined in Article 2 of the Convention establishing the World Intellectual Property Organization, done at Stockholm, 14 July 1 967;

(d) "Joint research" means research that is implemented with financial support from one or both Parties and that involves collaboration by participants from both the Community and the United States of America, and is designated as joint research in writing by the Parties or their scientific and technological organizations and agencies, or in the case where there is funding by only one Party, by that Party and the participants in that project;

(e) "Participants" means any individual or entity, including inter alia, the Parties' scientific and technological organizations and agencies, private persons, undertakings, research centers, universities, subsidiaries of European and U.S. entities, or any other form of legal entity involved in cooperative activities.

ARTICLE 3

Principles

Cooperative activities shall be conducted on the basis of the following principles:

(a) Mutual benefit based on an overall balance of advantages; (b) Reciprocal opportunities to engage in cooperative activities; (c) Equitable and fair treatment; (d) Timely exchange of information which may affect cooperative activities.

ARTICLE 4

Areas of cooperative activities

(a) Sectors for cooperative activities are:

environment (including climate research); biomedicine and health (including research on AIDS, infectious diseases and drug abuse); agriculture; fisheries science; engineering research; non-nuclear energy; natural resources; materials sciences and metrology; information and communication technologies; telematics; biotechnology; marine sciences and technology; social sciences research; transportation; science and technology policy, management, training and mobility of scientists;

(b) The Parties may modify this list upon recommendation by the Joint Consultative Group mentioned in Article 6, in accordance with procedures in force for each Party.

(c) The Parties may jointly pursue cooperative activities with third parties.

ARTICLE 5

Forms of cooperative activities

(a) Subject to applicable laws, regulations and policies, the Parties shall foster, to the fullest extent practicable, the involvement of participants in cooperative activities under this Agreement with a view to providing comparable opportunities for participation in their scientific and technological research and development activities.

(b) Cooperative activities may take the following forms:

1. coordinated research projects and joint research projects; 2. joint task forces; 3. joint studies; 4. joint organization of scientific seminars, conferences, symposia and workshops; 5. training of scientists and technical experts; 6. exchanges or sharing of equipment and materials; 7. visits and exchanges of scientists, engineers or other appropriate personnel; 8. exchanges of scientific and technological information as well as on practices, laws, regulations and programs relevant to cooperation under this Agreement.

Where appropriate, such cooperative activities shall take place pursuant to implementing arrangements concluded between the Parties' executive agents, or their scientific and technological organizations and agencies. These arrangements may describe the nature and the duration of cooperation for a specific area or purpose, treatment of intellectual property as provided for in the Annex, funding, allocation of costs, and other relevant matters.

ARTICLE 6

Coordination and Facilitation of Cooperative Activities

(a) The coordination and facilitation of cooperative activities under this Agreement shall be accomplished on behalf of the Government of the United States of America by the Department of State and on behalf of the Community by the European Commission, acting as Executive Agents.

(b) The Executive Agents shall establish a Joint Consultative Group (hereinafter referred to as the "JCG") for the oversight of scientific and technological cooperation under this Agreement. The JCG shall consist of a limited equal number of official representatives of each Party.

(c) The JCG may hold consultations on general science and technology issues; exchange information; establish task forces and working groups as appropriate; consult experts as appropriate and needed; and otherwise work to increase mutual understanding of the Parties' activities and programs related to science and technology.

(d) The functions of the JCG shall include:

1. overseeing and recommending activities under the Agreement; 2. making recommendations pursuant to Article 4 (b); 3. advising the Parties on ways to enhance cooperation consistent with the principles set out in this Agreement; 4. annually providing a report on the status and effectiveness of cooperation undertaken under this Agreement; 5. reviewing the efficient and effective functioning of the Agreement.

(e) The JCG shall meet annually, unless otherwise agreed by the Parties. Meetings should be held alternately in the Community and the United States of America. The JCG shall establish its own rules of procedure, subject to approval by the Parties.

(f) Decisions of the JCG shall be reached by consensus. Minutes, comprising a record of the decisions and principal points discussed, shall be taken at each meeting. These minutes shall be agreed upon by those persons selected from each side to jointly chair the meetings.

ARTICLE 7

Funding and Legal Considerations

(a) Cooperative activities shall be subject to the availability of appropriated funds and to the applicable laws and regulations, policies and programs of the Community and the United States of America.

(b) Each Party shall bear the costs of discharging its responsibilities under this Agreement, including costs of participation in meetings of the JCG. However, costs, other than those for travel and accommodation, which are directly associated with meetings of the JCG, shall be borne by the host Party.

ARTICLE 8

Entry of Personnel and Equipment

Each Party shall take all reasonable steps and use its best efforts, within applicable laws and regulations, to facilitate entry to and exit from its territory of persons, material, data and equipment involved in or used in cooperative activities under this Agreement.

ARTICLE 9

Treatment of Intellectual Property

The allocation and protection of intellectual property rights under this Agreement shall be in accordance with the provisions of the Annex, which forms an integral part of this Agreement.

ARTICLE 10

Other Agreements and Transitional Provisions

(a) The Parties shall endeavor, where appropriate, to bring under the terms of this Agreement new arrangements for scientific and technological cooperation between the Community and the Government of the United States of America that fall under the scope of Article 4.

(b) This Agreement is without prejudice to rights and obligations under other agreements between the Parties and any agreement or arrangement between either of the Parties and non-participant third parties, including agreements or arrangements between their scientific and technological organizations or agencies and a Member State of the Community.

ARTICLE 11

Territorial Application

This Agreement shall apply, on the one hand to the territories in which the Treaty establishing the European Community is applied and under the conditions laid down in that Treaty, and on the other hand to the territory of the United States of America. This shall not prevent the conduct of cooperative activities on the high seas, outer space, or the territory of third countries, in accordance with international law.

ARTICLE 12

Entry into Force, Termination and Dispute Settlement

(a) This Agreement shall enter into force on the date on which the Parties have notified each other in writing that their respective internal procedures necessary for its entry into force have been completed.

(b) This Agreement is concluded for an initial period of five years. Subject to review by the Parties in the final year of each successive period, the Agreement may be extended, with possible amendments, thereafter for additional periods of five years by mutual written agreement between the Parties.

(c) This Agreement may be terminated at any time by either Party upon six months' written notice. The expiration or termination of this Agreement shall not affect the validity or duration of any arrangements made under it, or any specific rights and obligations that have accrued in compliance with the Annex.

(d) This Agreement may be amended by agreement of the Parties. Amendments shall enter into force on the date on which the Parties have notified each other in writing that their respective internal procedures necessary for amending this Agreement have been completed.

(e) All questions or disputes related to the interpretation or implementation of this Agreement shall be settled by mutual agreement of the Parties.

ARTICLE 13

This Agreement is signed in duplicate in the Danish, Dutch, English, Finnish, French, German, Greek, Italian, Portuguese, Spanish, and Swedish languages, each of these texts being equally authentic.

ANNEX — INTELLECTUAL PROPERTY

Pursuant to Article 9 of this Agreement;

The Parties shall ensure adequate and effective protection of intellectual property created or furnished under this Agreement and relevant implementing arrangements. The Parties agree to notify one another in a timely fashion of any inventions or copyrighted works arising under this Agreement and to seek protection for such intellectual property in a timely fashion. Rights to such intellectual property shall be allocated as provided In this Annex.

I. SCOPE

A. This Annex is applicable to all cooperative activities undertaken by the Parties or their participants pursuant to this Agreement, except as otherwise specifically agreed by the Parties.

B. For purposes of this Agreement, "intellectual property" shall have the meaning found in Article 2 of the Convention Establishing the World Intellectual Property Organization, done at Stockholm, July 14, 1967.

C. This Annex addresses the allocation of rights, interests, and~ royalties between the Parties or their participants. Each Party shall ensure that the other Party or its participants can obtain the rights to intellectual property allocated in accordance with the Annex. This Annex does not otherwise alter or prejudice the allocation between a Party and its nationals, which shall be determined by that Party's laws and practices.

D. Disputes concerning intellectual property arising under this Agreement should be resolved through discussions between the relevant participants, or, if necessary, the Parties. Upon mutual agreement of the Parties, the participants may submit a dispute to an arbitral tribunal for binding arbitration. Unless the participants agree otherwise in writing, the arbitration rules of UNCITRAL shall govern.

E. Termination or expiration of this Agreement shall not affect rights or obligations under this Annex.

II. ALLOCATION OF RIGHTS

A. Each Party shall be entitled to a non-exclusive, irrevocable, royalty-free license in all countries to reproduce, publicly distribute and translate scientific and technical journal articles, non-proprietary scientific reports, and books directly arising from cooperation under this Agreement. All publicly distributed copies of a copyrighted work prepared under this provision shall indicate the names of the authors of the work unless an author explicitly declines to be named. Each Party or its participants shall have the right to review a translation prior to public distribution.

B. Rights to all forms of intellectual property, other than those rights described in paragraph II (A) above, shall be allocated as follows:

1. Visiting researchers, for example, scientists visiting primarily in furtherance of their education, shall receive intellectual property rights under arrangements with their host institutions. In addition, each visiting researcher named as an inventor shall be entitled to treatment as a national of the host country with regard to awards, bonuses, benefits, or any other rewards, in accordance with the policies of the host institution.

2. (a) For intellectual property which is or may be created during joint research, the Parties or their participants shall jointly develop a technology management plan. The technology management plan shall consider the relative contributions of the Parties and their participants, the benefits of licensing by territory or for fields of use, requirements imposed by the Parties' domestic laws, and other factors deemed appropriate.

(b) If the parties or their participants did not agree to a joint technology management plan in the initial research cooperation agreement and cannot reach an agreement within a reasonable time, not to exceed six months, from the time a Party becomes aware of the creation or likely creation of the intellectual property in question as a result of the joint research, the Parties or their participants shall resolve the matter in accordance with the provisions of paragraph I (D). Pending resolution of the matter, such intellectual property shall be owned jointly by the Parties or their participants, but shall be commercially exploited (including product development) only by mutual agreement.

(c) "Joint research" means research that is implemented with financial support from one or both Parties and that involves collaboration by participants from both the Community and the United States of America and is designated as joint research in writing by the Parties or their scientific and technological organizations and agencies, or in the case where there is funding by only one Party, by that Party and the participants in that project.

(d) In the event that either Party believes that a particular joint research project under this Agreement has led or will lead to the creation or furnishing of a type of intellectual property that it protects but is not protected throughout the territory of the other Party, the Parties shall immediately hold discussions to determine the allocation of the rights to the said intellectual property. The Joint activities in question will be suspended during the discussions, unless otherwise agreed by the Parties thereto. If no agreement can be reached within a three month period from the date of the request for discussions, cooperation on the project in question will be suspended or terminated at the request of either Party.

III. PROPRIETARY INFORMATION

In the event that information identified in a timely fashion as proprietary is furnished or created under the Agreement, each Party and its participants shall protect such information in accordance with applicable laws, regulations, and administrative practice. Without prior written consent, none of the Parties shall disclose any proprietary information except to employees, government personnel, and prime and subcontractors. Such disclosures shall be for use only within the terms of their permits or licenses with the Parties or the scope of work of their contracts with the Parties and in work relating to the subject matter of the information so disseminated. The Parties shall impose, or shall have imposed, through appropriate arrangements such as research contracts, grant documents, technology management plans, etc. an obligation on all participants receiving such information to keep it confidential.

If one of the Parties becomes aware that, under its laws or regulations, it will be, or may reasonably be expected to become, unable to meet the non-disclosure provisions, it shall immediately inform the other Party. The Parties shall thereafter consult to define an appropriate course of action. Information may be identified as proprietary if it is secret in the sense that it is not, as a body or in the precise configuration or assembly of its components, generally known or readily accessible by lawful means; has actual or potential commercial value by virtue of its secrecy; has been subject to steps that were reasonable under the circumstances by the person lawfully in control, to maintain its secrecy; and not already in the possession of the recipient without an obligation concerning its confidentiality.

Conference Participants

Philip Abelson
American Association for the
Advancement of Science

Jean-Francois Abramatic
World Wide Web Consortium

Enrico Alleva
Instituto Superiore di Sanita, Italy

Jane Alspach
American Association of Engineering
Societies

Pablo Amor
U.S. Delegation of the European
Union

Kiyoshi Ando
Nikkei

Anders Backlund
Kvaerner Masa-Marine

Ouahid Bakouche
Embassy of France

Dennis Baldocchi
National Oceanic and Atmospheric
Administration

Elizabeth Baldwin
Optical Society of America

Wendy Baldwin
National Institutes of Health

Sharon Bank
Transportation Research Board

Thomas Barnwell
National Exposure Research
Laboratory

Jon Baron
Department of Defense

David Beckler
Carnegie Commission

PARTICIPANTS

Ed Behrens
Procter and Gamble

Johannes Belz
German-American Academic Council

Dorothy Bergamaschi
Department of State

Ake Bergman
Stockholm University

Francois Bertin
Embassy of France

Robert J. Betsold
Turner Fairbank Highway Research Center

Amar Bhat
National Institutes of Health

Richard Bissell
National Research Council

Richard Biter
Department of Transportation

Dr. Blanc
GEC Alsthom Signalling

Wilhelmus Blonk
European Commission, DGVII

Joseph Bordogna
National Science Foundation

Michael Borrus
University of California, Berkeley

Brad Botwin
Department of Commerce

Sandor Boyson
University of Maryland

Jeffrey Brancato
National Science Foundation

Rick Brennan
GE Information Services

Mel Briscoe
United States Navy

Richard Brook
Engineering & Physical Sciences Research Council, UK

Abraham Brouwer
Wageninigen Agricultural University, The Netherlands

Steven Buchsman
Department of State

John Cadogan CBE, FRS
Research Councils, UK

Joseph F. Canny
Department of Transportation

Giulio Cantarella
University of Reggio Calabria, Italy

Fenton Carey
Department of Transportation

Steve Carpenter
National Institute of Standards and Technology

Ian Carter
University of Glasgow

Paul Cederborg
National Academy Press

Roberto Cencioni
European Commission, DGXIII

Mel Ciment
National Science Foundation

Steve Clemons
Economic Strategy Institute

Iain Cockburn
University of British Columbia

Jerry Cogan
Milliken Research

Tom Coleman
BASF

E. William Colglazier
National Research Council

T. D. Collinsworth
Department of Defense

Sara Comley
International Observers

Anna Constantinidou
Embassy of Greece

Henri Conze
Ministry of Defense, France

Stephen L. Cooney
Siemens Corporation

Robert W. Corell
National Science Foundation

Lucio G. Costa
University of Washington

George Counts
US-EU Task Force on Communicable Diseases, NIH

Kelley S. Coyner
Department of Transportation

Wolfgang Cramer
Potsdam-Institut fur Klimafolgenforschung, Germany

John C. Crawford
Sandia National Laboratories

Mark H. Crawford
New Technology Week

Michael Crow
Columbia University

Ulrich Cubasch
Deutsches Klimarechenzentrum GmbH, Germany

Christina Curtin
Environmental Science and Technology

George Daston
Procter & Gamble

Mike Davey
Congressional Research Service

Fonseca de Moura
Carnegie Mellon University

Jim DeCorpo
United States Navy

PARTICIPANTS

Raoul Delcorde
Embassy of Belgium

Papken S. Der Torossian
Silicon Valley Group

Maxx Dilley
US Agency for International Development

Gerry Dinneen
National Research Council

Dee Ann Divis
GPS World Magazine

Bruce Don
Critical Technologies Institute

Mortimer Downey
Department of Transportation

J.C. Duplessy
France

Robert Eagan
Sandia National Laboratories

Paul Eckert
Office of Senator John Breaux

Martin Eichtinger
Embassy of Austria

Stuart E. Eizenstat
Department of State

Frank Elfring
United States Coast Guard

Maria Eli
The European Institute

Margarete Endl
Journalist

Stephen Eule
House Science Committee

Jean-Pierre Euzen
European Commission

Carey Fagan
Federal Aviation Administration

Ana Faisca
Ministry of Science and Technology, Portugal

Paolo Fasella
Director General for Research, Italy

Penelope A. Fenner-Crisp
Environmental Protection Agency

Alda Fernandes
Embassy of Portugal

Peter Finnerty
Sea-Land Service, Inc.

Frank Finver
Department of State

Kenneth Flamm
Brookings Institution

Christian Fluhr
Conseiller du Directeur de la DIST, France

Gioacchino Fonti
MURST

Joshua Foster
National Oceanic and Atmospheric
Administration

Paul Foster
Chemical Industry Institute of
Toxicology

Robert Frederking
Carnegie Mellon University

Peter Fritz
UFZ Umwaltforachungszentrum,
Germany

Bill Frymoyer
Office of House Democratic Leader
Richard Gephardt

Henry Fuchs
University of North Carolina,
Chapel Hill

Irene Gabriel
Federal Ministry of Science and
Transport, Austria

Asa Gahne
Embassy of Sweden

Carmen Garcia
Embassy of Spain

Peter Gelbke
BASF AG, Germany

Rainer Gerold
European Commission, DGXII

Leslie A. Gerson
Department of State

Richard Getzinger
American Association for the
Advancement of Science

Anver Ghazi
European Commission, DG XII

John Giesy
University of Michigan

Joe Giglio
Northeastern University

Jim Glass
Massachusetts Institute of Technology

H. Glatz
Trans-Atlantic Business Dialogue

Francois Govaerts
European Commission, DG XII

Martin Grabert
KOWI, Germany

Nicholas E. Graham
International Research Institute

Thomas Grandke
Siemens Corporate Research, Inc.

Jacqueline Grapin
The European Institute

L. Earl Gray
National Health & Environmental
Effects Research Laboratory

Martha Graybowski
Renesselear Polytechnic Institute

Dan Greenberg
Science & Government Report

John Gresham
DDR&E, Department of Defense

Jeff Grove
House Science Committee

Eva Guterres
Embassy of Sweden

Erik Habers
European Commission, DGIII

Matthias Hack
Bundesministerium fur Bildun und
Wissenschaft, Germany

Herbert Hager
University of Agricultural Sciences,
Austria

Manuel Hallen
National Science Foundation

Jack Halpern
University of Chicago

Robert L. Hance
Motorola

Aaron Hand
Photonics Spectra

Gerald Hane
Office of Science and Technology
Policy

Donna K. Harman
National Institute of Standards and
Technology

Dan Hartley
Sandia National Laboratories

Juris Hartmanis
National Science Foundation

Dr. Hasselman
Max-Plank Institut, Germany

Yvon Heckscher
Heckscher Professional Group

Susan Hedigan
University College Dublin, Ireland

Maria Hedqvist
Embassy of Sweden

Colin Helmer
Department of State

Bill Hendrickson
Issues in Science and Technology

Bert Herzog
Computer Graphics, Inc.

Thuy Hia
Department of Commerce

Lynette Hirschman
Mitre

Ron Hodge
General Electric

Alice Hogan
Office of Science and Technology
Policy

Charles Holland
Department of Defense

Brooke Holmes
Department of State

John B. Horrigan
National Research Council

John C. Horsley
Department of Transportation

Manfred Horvat
BIT, Austria

Edward Howard
National Oceanic and Atmospheric Administration

Kay Howell
NCOCIC

Jeanne Hudson
National Science Foundation

Caitlin Hughes
Department of Transportation

Charles A. Hunnicutt
Department of Transportation

Sharon Hyrnkow
National Institutes of Health

Veijo Ilmavirta
Helsinki University of Technology, Finland

Marie-Christine Imbert
INRIA, France

Richard Jackson
National Institute of Standards and Technology

Said Jahanmir
Department of Commerce

William James
Procter and Gamble

Margaret Jenny
U.S. Airways

James Jensen
National Academy of Sciences

Lynn Johnson
National Academy of Science

Peter Jones
Transport Studies Group, UK

Gary Jones
Sandia National Laboratories

Gilbert Kalb
GMD German National Research Center for Information Technology

Tom Kalil
National Economic Council
The White House

Ray Kammer
National Institute of Standards and Technology

Marie-Ange Katzeff
Embassy of Belgium

Robert J. Kavlock
National Health & Environmental Effects Research Laboratory

Samuel Kavruck
Washington Counseletter

Martin Kayser
BASF

PARTICIPANTS

Keith Keen
European Commission, DGVII

Hannu Kemppainen
Tekes, Finland

Melinda L. Kimble
Department of State

Kelly Kirkpatrick
Office of Science and Technology Policy

Judith Klavans
Columbia University

John P. Klus
University of Wisconsin-Madison

Martin Koubek
Department of Transportation

Steve Krauwer
University of Utrecht, The Netherlands

John Krebs, FRS
Natural Environment Research Council, UK

Norman Kreisman
Department of Energy

Anssi Kujala
Embassy of Finland

Damian Kulash
ENO Foundation

Kathleen Kunzer
Chemical Manufacturers Association

Kristina A. Kvien
Department of State

Patrice Laget
US Delegation of the European Union

Gordon John Lake
European Parliament

Richard Lambert
Department of Health and Human Services

Ron Larsen
Defense Advanced Research Projects Agency

Lisbeth Lawrence
United Medical & Dental Schools

Graham Lawton
Chemistry and Industry Magazine

Carolyn Leep
Chemical Manufacturers Association

Hans Lehmann
Kontakstelle Biomed, Germany

Risto Lemmelä
Helsinki University of Technology, Finland

Wil Lepkowski
Chemical & Engineering News

Josh Lerner
Harvard Business School

Michael Lesk
National Science Foundation

Tore Li
Royal Norwegian Embassy

Helmut List
Industrial Research and Development
Advisory Council

Ron Lorton
Department of State

George Lucier
National Institute of Environmental
Health Sciences

George Luckett
Shell Chemical Europe Ltd.

Janet Lynch
General Electric

Johannes Maier
Bosch

Erminio Marafante
Ispra

Gennaro Marino
University of Naples, Italy

Steve Mautner
National Academy Press

Gail McCarthy
Electric Power Research Institute

Roger McClellan
Chemical Industry Institute of
Toxicology

Bill McCluskey
United States Navy

Clark McFadden
Dewey Ballantine

Jean-Pierre Medevielle
INRETS, France

Joaquin Melia
Universitat de València, Spain

Jose Amaral Mendes
University of Evora, Portugal

Steve Merrill
National Research Council

Gerard Meyer
Carnegie Mellon Research Institute

John C. Miles
Ankerbold International Ltd.

Kevin Mills
Defense Advanced Research Projects
Agency

Norman Y. Mineta
Lockheed Martin IMS

Ana Mirones
Portugal

Alfonso Molina
University of Edinburgh

Michael Moloney
Embassy of Ireland

Linda Moodie
National Oceanic and Atmospheric
Administration

Duncan T. Moore
Office of Science and Technology Policy

Gordon Moore
Intel Corp.

Bill Morin
R. Wayne Sayer & Associates

Grant Moser
Business Publishers

Kelly Jacobs Mudd
Environmental Protection Agency

Mort Mullins
Chemical Manufacturers Association

Jeremiah Murphy
Siemens Corporation

Antonio Navarra
Consiglio Nazionale delle Ricerche, Italy

Eric A. Nerlinger
Zentrum fur Europaische Wirtschaftsforschung, Germany

Mikko Niini
Kvaerner Masa-Yards, Finland

Stefan Noll
Fraunhofer-Institut fur Graphische, Germany

Robert C. North
United States Coast Guard Headquarters

Robert Norwood
NASA Headquarters

W.C. Oechel
San Diego State University

John C. Oldfield
National Research Council

Scott Pace
Critical Technologies Institute, RAND

Hugo Paemen
U.S. Delegation of the European Union

Jeff Paniati
Department of Transportation

Ron Parsons
CommerceNet

Markus Pasterk
Federal Ministry for Science and Transport, Austria

Marcus Pattloch
DFN-Verein, Germany

Louis-Francois Pau
Ericsson, Sweden

Fabian Pease
Defense Advanced Research Projects Agency

John Sarborg Pedersen
Embassy of Denmark

Maria Luz Penacoba
Spain

Karin Petersen
Palo Alto Research Park

William A. Peterson
Department of Education

Kees Planqué
Embassy of The Netherlands

Gary Poehlein
National Science Foundation

Alan Poole
DOW Europe S.A.

E. Praestgaard
European Science and Technology Assembly

Peter Preuss
Environmental Protection Agency

Thomas Price
American Association of Engineering Societies

William E. Primosch
Department of State

Knud Prytz
Scandlines, Denmark

George Radda
Medical Research Council, UK

F.J. Radermacher
FAW Ulm, Germany

Saifur Rahman
National Science Foundation

Geoff Randall
Zeneca, UK

Brian Randell
University of Newcastle, UK

Steve Rattien
RAND

Scott Rayder
House Science Committee

Ruth Reck
University of California, Davis

Lucy H. Richards
Department of Commerce

Giovanni Rinaldi
Italy

John Rodman
RAMS-FIE

Philippa Rogers
Embassy of Great Britain

Laura Rosato
L.R. Associates

Ronald Rosenfeld
Carnegie Mellon

Christopher Ross
US Delegation of the European Union

M.D.A. Rounsevell
Cranfield University, UK

Jorma Routti
European Commission, DGXII

Tom Rozzell
National Research Council

Roland Ruhle
University of Stuttgart, Germany

Jason Rushton
Innolog

Lee Sanders
University of Warwick

Scott Sandgathe
United States Navy

Margarida Santos
Instituto de Cooperacao Cientifica e
Tecnologica Internacional, Portugal

Roger Sattler
University of Maryland

Claire Saundry
National Institute of Standards and
Technology

R. Wayne Sayer
R. Wayne Sayer & Associates

Wendy Schacht
Congressional Research Service

Dr. Schacke
Ministry of Transportation,
Denmark

Ingolf Schaedler
Federal Ministry for Science and
Transport, Austria

Wolfgang Schlump
Embassy of Germany

Wolfram Schoett
Embassy of Germany

Christopher Schonwalder
National Institute of Environmental
Health Sciences

Gregory Schuckman
American Association of Engineering
Societies

Craig Schultz
National Research Council

Stuart J.D. Schwartzstein
Office of Naval Research Europe

Bernard Schwetz
Food and Drug Administration

Rob Scott
Economic Policy Institute

Alan Sears
Defense Advanced Research Projects
Agency

H. Segner
Center for Environmental Research,
Germany

Dieter Seitzer
Fraunhofer Institute, Germany

Lisa Shaffer
University of California

John Shamaly
Silicon Valley Group

Michael Shelby
NIEHS

Kenneth Shine
Institute of Medicine

Jagadish Shukla
Institute of Global Environment and
Society, Inc.

Susan M. Sieber
Division of Cancer Epidemiology

Robert Skinner
Transportation Research Board

Horst Soboll
DaimlerBenz Technology, Germany

Randall Soderquist
Office of Senator Bingaman

Elizabeth Sokul
House Committee on Science

Micheal Sollosi
US Coast Guard

Bill Spencer
SEMATECH

Scott Stafford
DDR&E, Department of Defense

Linda Staheli
National Institutes of Health

Attilio Stajano
European Commission, DG III

Volker Steinbiss
Philips Speech Processing, Germany

Robert Stern
Consultant in Technology Management

Carrie Stevens
U.S. General Accounting Office

Macol Stewart
National Oceanic and Atmospheric Administration

Matthew Stiff
Museum Documentation Association, United Kingdom

Deborah Stirling
Stirling Strategic Services

George Strawn
National Science Foundation

Gary W. Strong
National Science Foundation

Orson Swindle
Federal Trade Commission

Istvan Szemenyei
Embassy of the Republic of Hungary

Rob Taalman
CEFIC-EMSG

Tyrone Taylor
Federal Laboratory Consortium

Alexander Tenenbaum
Embassy of Italy

Francois-Xavier Testard-Vaillant
Embassy of France

William B. Testerman, II
House Science Committee

Costantino Thanos
CNR, Italy

Richard Thayer
Telecommunications & Technologies International

Bonnie H. Thompson
National Science Foundation

Gavin Thomson
European Economic Development Services, Ltd

Sandra Tirey
Chemical Manufacturers Association

Marja-Leena Tolonen
TEKES, Finland

Alan Tonelson
US Business and Industrial Council Educational Foundation

Francoise Touraine-Moulin
Embassy of France

Kevin Trenberth
National Center for Atmospheric Research

Charles Trimble
Trimble Navigation

Robert Tuch
German-American Academic Council

James Turner
House Science Committee

Franklin Urteaga
Office of Science and Technology Policy

Steve Usdin
Endocrine-Estrogen Newsletter

Hans Uszkoreit
DFKI, Germany

Allie Uyehara
Uyehara International Associates Inc.

Cecil Uyehara
Uyehara International Associates Inc.

Riccardo Valentini
Universita della Tusscia, Italy

Anja Van Dam
Royal Netherlands Embassy

Thierry van der Pyl
European Commission, DGIII

Reinder J. Van Duinen
European Science and Technology Assembly

G. van Oortmerssen
CWI, The Netherlands

Gerrit Vanderwees
Embassy of the Netherlands

Dave Varney
Federal Information and News Dispatch

Pedro Veiga
FCCN, Portugal

Nicholas Vonortas
George Washington University

J.G. Vos
RIVM, The Netherlands

Meredith Wadman
Press

Caroline Wagner
Rand

Shukri Wakid
Department of Commerce

Michael Wallace
University of Washington

Michael Waters
National Health and Environmental
Effects Research Laboratory

Albert Wavering
National Institute of Standards and
Technology

Peter Webster
University of Colorado

Tom Weimer
National Academy of Sciences

Clifford Weinstein
Massachusetts Institute of Technology

Charles Wessner
National Research Council

John Westensee
Aarhus School of Business, Denmark

Christopher Whaley
Embassy of Great Britain

Chelsea C. White
University of Michigan

Wendy White
National Academy of Sciences

Isabel Wolte
Embassy of Austria

David N. Wormley
Pennsylvania State University

William Wulf
National Academy of Engineering

Kenneth Wykle
Department of Transportation

Dr. Yelloz
GEC Alsthom Signalling

Antonio Zampolli
Instituto di Linguistica
Computazionale, Italy